Environmental Hazards

Series editor
Thomas A. Birkland

More information about this series at http://www.springer.com/series/8583

Asmita Tiwari

The Capacity Crisis in Disaster Risk Management

Why disaster management capacity remains low in developing countries and what can be done

Springer

Asmita Tiwari
Rockefeller College of Public
Administration and Policy
University at Albany
Albany, NY, USA

All views in this book are author's and do not represent that of her current or past employers.

The Author acknowledges that there may be instances where she was unable to trace or contact the copyright holder for permission to reproduce/adapt selected material in this volume. The Author has included complete source references for all such material and takes full responsibility for these matters. If notified, the publisher will be pleased to rectify any errors or omissions at the earliest opportunity.

Environmental Hazards
ISBN 978-3-319-09404-5 ISBN 978-3-319-09405-2 (eBook)
DOI 10.1007/978-3-319-09405-2

Library of Congress Control Number: 2014958901

Springer Cham Heidelberg New York Dordrecht London

Printed on acid-free paper

Springer International Publishing AG Switzerland is part of Springer Science+Business Media (www.springer.com)

Preface

Imagine a scenario. A powerful earthquake shook many cities and rural areas of a developing country. Thousands of people are killed—trapped in their houses, offices, schools, hospitals, and shops, which have been reduced to rubbles. A majority of citizens are traumatized after losing their loved ones and entire life's assets. Massive aid flows in. National and state governments, charity organizations, private sector, citizens, international and national aid agencies donate generously in the aftermath of this dreadful disaster. Apart from relief and recovery, the government promises to use this aid for long-term 'capacity building'—to prevent and reduce future disaster impacts.

Now think. If you revisit one of the affected cities 10 years after the earthquake, what are you most likely to find? You would imagine that the people in this place are better prepared for an imminent disaster since they sit on an earthquake fault. You would think that the massive aid that was collected for 'capacity building' was put to work and the new buildings and infrastructure are earthquake-resistant. You will imagine that the reconstructed cities, which were almost totally destroyed in the last earthquake, have wider roads for emergency vehicles, with alarms and drills to alert and prepare people. An emergency crew is ready to respond, relief materials and emergency funding are quickly available.

However, far from the expected findings, you find that citizens are no better prepared, no more concerned about their own safety. Many buildings and infrastructure are likely to fall down, even in low-intensity quakes. While there might be some means to alert people and rescue/relief materials stored in identified shelters, it is not enough. Higher levels of governments (national and state levels) believe that they have invested in preventing future impacts but are not taking any further actions. Local governments, private sector, non-profit charity organizations, and nongovernmental organizations know that people are not prepared and more is needed. But they are not taking any steps. If there is another earthquake in the city, there will be a lot more damages and fatalities this time—given that population and assets have grown. This is a 'capabilities trap' situation, where even after conscious capacity development efforts by the government and donors, there is no effective capacity.

One might think that the above scenario is hypothetical. In fact, this scenario is not only real but also very common. A number of deadly disasters hit the world in recent years including a cyclone in Philippines (2012) killing 1,100 people, earthquake in Indonesia (2009) killing over 1,000 people, cyclone Nargis in Myanmar (2008) killing 138,366 people, and the Sichuan earthquake in China (2008) killing 90,000 people. The 2010 earthquake in Haiti alone caused more than 200,000 deaths. Most of these places had suffered from earlier disasters.

With increased global attention to disaster prevention and preparedness, many developing countries are undertaking capacity building programs to reduce the extent of damages from disasters. The disaster mitigation activities include preventive actions aimed at reducing loss of lives and damages resulting from disasters. Governments and donors now agree that it is more cost-effective to invest in disaster management and preparedness activities than to keep on providing major relief and reconstruction aid. However, disasters continue to strike causing higher damages and losses, especially in areas that have witnessed disasters in the past. Overall, disaster events as well as number of affected people and disaster damages are increasing globally. Between January 1975 and October 2008, the international emergency disasters database (EM-DAT) recorded 8,866 natural disaster events killing more than two billion people with the majority of related mortality and losses (relative to GDP) concentrated in the last two decades and in low- and middle-income countries.

How can a place be built and managed so that it is safe for people to live? Ironically, many governments and people keep on asking the same question after every new disaster. Why, even with a high level of investment in increasing government's capacity to manage disasters, do the impacts of disasters continue to increase? Is capacity development in managing disasters not working? This book is about answering these questions, highlighting how current capacity development efforts for managing disasters are leading to capacity crisis or a capability trap situation. However, the main point of the book is not of a doomsday prediction—to sound alarm about more failures and higher disaster impacts in developing countries. On the contrary, this book is primarily about hope, optimism, and change. The book provides an alternative and a better way to develop effective capacity for preventing and managing future disaster impacts.

The solutions to the questions raised are based on two main lines of enquiries. First, what capacities are actually needed, and second, how to develop and sustain such capacities to ensure that they are effective in the long run. The enquiries are based on an assessment of current literature in international capacity development and disaster risk management fields, and an in-depth case study in three earthquake-affected towns of Gujarat, India, relying on interviews and surveys. A comparison of countries with better disaster prevention and response capacities with those that are not able to do so is also undertaken to support recommendations.

The book is divided into two main parts. The first part, comprising of first two chapters, will set the scene on how there is a capacity crisis for managing disasters, particularly in developing countries. This part, deriving from the case study of capacity development after the 2001 earthquake in Gujarat, India, will raise critical questions on the meaning of capacity crisis situation. The second forward looking part, comprising of the remaining six chapters, will focus on what is meant by sustainable and effective capacity for managing and mitigating disaster impacts. Relying on current literature, field research, practical insights, and experiences of other countries, this part will provide recommendations for Gujarat case and general implications for donors, governments, and communities.

Albany, NY, USA Asmita Tiwari

Acknowledgements

This book is a culmination of an extensive research carried out as a part of my doctoral study. My passion for the field, however, started a long time ago—back in 1984 with the Union Carbide's leakage of methyl isocyanate gas (known as "Bhopal Gas Tragedy") in Bhopal, my hometown in India. Images of thousands of people screaming and running to escape the gas, which was causing blindness and shortness of breath are fixed in my mind. Was there a way to avoid such a tragedy? I often asked myself as I went back to school in Bhopal after several months long vacation after the 'tragedy'. Later, while I was working as a public management professional in Gujarat, I witnessed the Kutch earthquake in 2000. This time it was a larger scale tragedy—with numerous deaths and victims. But the images of people running in confusion and panic were similar to the 1984 Bhopal Gas Tragedy. Is there no way to escape disasters? Can we at least be prepared for the aftermath? In my quest, I was lucky to work in a United Nations agency focusing on drought response, and later in another international development agency as a Disaster Risk Management Specialist. In the later position, I was able to witness many more post-disaster situations around the world. While my journey still continues, this book presents the questions I raised along the way and the answers I discovered.

Several remarkable people have made it possible for me to complete this work. First of all, I am grateful to the people of Kutch, Gujarat, and several other places across the world, whose enduring spirit has a central role in the development of concepts in this book. I am thankful to my dissertation chair Professor Jeffrey Straussman who saved me from my own capability trap and provided an overall direction to my research. I am indebted to Professors Sue Faerman and Ray Bromley for their timely guidance and tireless reviews. I am thankful to Thomas Birkland, series editor for *Environmental Hazards*, for his guidance, encouragement, and careful review of the manuscript. I am thankful to Yogesh Jadeja, Director of Arid Communities and Technologies (ACT), and his team as well as my pal Utkarsh Patel for their immense help during my field work. I am thankful to my employers, current and past, for the experiences, which gave me the broader understanding of different issues across the world.

I am grateful to my parents Mahesh and Shobha Tiwari for strongly supporting and encouraging me to be my best. I am extremely thankful to my husband Sree Nampoothiri for being the best mentor, critic, and friend. His tireless encouragement and support made it possible for me to complete this work in time. Finally, I dedicate this book to my sons Shreyas and Shourya as well as to children all over the world whose future depends a lot on how well we manage the risks in our world.

Contents

About the Author

Asmita Tiwari is an international development practitioner in the fields of disaster risk management, urban development, climate change, and environmental management. Growing up in Bhopal, India, she was deeply affected by the Union Carbide disaster (also known as Bhopal Gas Tragedy) that took lives of thousands of people in one night. Her passion to understand more about disaster management took her to multiple countries in Africa, the Middle East, South Asia, and North America—while working for major international development agencies as Disaster Risk Management professional. Her primary task has been to support governments in the aftermath of disasters, particularly in developing their capacity to reduce future disaster risks and impacts. This book is a culmination of her fifteen-year long journey to understand how to better manage disasters and disaster risks. She has a Ph.D. in Public Administration and Policy, and Master's in Urban and Regional Planning.

Abbreviations

ACT	Arid Communities and Technologies
ADA	Area Development Authority
ADB	Asian Development Bank
ADDO	Additional District Development Officer
BADA	Bhachau Area Development Authority
BDC	Bhuj Development Council
BHADA	Bhuj Area Development Authority
BJP	Bharatiya Janata Party
CBO	Community-Based Organization
CEO	Chief Executive Officer
CEPT	Center for Environmental Planning and Technology
CIDA	Canadian International Development Agency
DDO	District Development Officer
DEOC	District Emergency Operations Center
DM	District Magistrate
DRM	Disaster Risk Management
EM-DAT	Emergency Database
EU	European Union
GDP	Gross Domestic Product
GEERP	Gujarat Emergency Earthquake Reconstruction Project
GIDM	Gujarat Institute of Disaster Management
GSDMA	Gujarat State Disaster Management Authority
GTZ	Deutsche Gesellschaft für Technische Zusammenarbeit
Ha	Hectare (1 Ha = 10,000 sq meter or 12,000 sq yard)
HFA	Hyogo Framework for Action
HIG	High-Income Group
IAS	Indian Administrative Service
IFAD	International Fund for Agriculture Development
INC	Indian National Congress
INEE	Inter-Agency Network for Education in Emergencies

JICA	Japanese International Cooperation Agency
KMVS	Kutch Mahila Vikas Sanghathan
KNNA	Kutch Nav Nirman Abhiyan
KPMG	Klynveld Peat Marwick Goerdeler
LIG	Low-Income Group
MIG	Middle-Income Group
NDMA	National Disaster Management Authority
NGO	Non-government Organization
NIDM	National Institute of Disaster Management
OECD	Organization for Economic Co-operation and Development
OED	Operations Evaluation Department
PRI	Panchayati Raj Institutions
Rs.	Indian Rupees
SEOC	State Emergency Operations Center
TDO	Taluka Development Officer
UN	United Nations
UNDP	United Nations Development Program
UNESCO	United Nations Educational, Scientific, and Cultural Organization
UNISDR	United Nations International Strategy for Disaster Risk Reduction
USA	United States of America
USAID	United States Agency for International Development
USD	United State Dollar
VRTI	Vivekananda Research and Training Institute
WFP	World Food Program

Part I
Capacity Crisis

Chapter 1
Disasters and Capacity Development Mirage

Abstract Disaster impacts are increasing globally. At the same time, disaster risk management capacity remains low in many developing countries despite conscious capacity development efforts. Apart from human agony and pain, this *capability trap* situation affects a country's economic and social development potential, pushing it into a lagged developmental trajectory. This chapter introduces the central theme of this book, illustrating with examples how some developing countries are facing capability trap situation in spite of conscious capacity development efforts by governments and international development agencies. The chapter starts with an overview of rising disaster impacts worldwide followed by the kind of capacity development programs usually undertaken to prevent and prepare for disasters and how these programs are mostly unable to develop effective capacity. With examples from various countries facing recurrent disasters and rising disaster impacts, this chapter raises many critical questions to discuss challenges associated with capacity development in disaster risk management, primarily in a developing country context. The chapter provides an overview of how the book is organized and how solutions to various questions, which are raised in this chapter, are explored in the remaining book.

Keywords Capacity development • Capability trap • Developing countries • Disaster impacts • Disaster risk management

Disaster impacts are increasing globally. At the same time, disaster risk management capacity remains low in many developing countries despite conscious capacity development efforts. Apart from human agony and pain, this *capability trap* situation affects a country's economic and social development potential, pushing it into a lagged developmental trajectory. Why are some countries facing capability trap situation? Is it because disasters are unique and pose a challenge that cannot be effectively managed by human societies? Is it dependent on a country's unique and intrinsic factors—such as economic well-being? Or is it because of the way capacity development for disaster risk management is taking place—in other words the design and implementation of capacity building programs are not good enough? What can be done to break the *capability trap*?

A. Tiwari, *The Capacity Crisis in Disaster Risk Management*, Environmental Hazards, DOI 10.1007/978-3-319-09405-2_1

This chapter introduces the central theme of this book, illustrating with examples how some developing countries are facing "capability trap" situation in spite of conscious capacity development efforts by governments and international development agencies. The chapter starts with an overview of rising disaster impacts worldwide followed by the kind of capacity development programs usually undertaken to prevent and prepare for disasters and how these programs are mostly not able to develop effective capacity. With examples of various countries facing recurrent disasters and rising disaster impacts, this chapter raises many critical questions to discuss challenges associated with capacity development in disaster risk management, primarily in a developing country context. Toward the end, the chapter provides an overview of how the book is organized and how solutions to various questions, which are raised in this chapter, are explored in the remaining book.

1.1 Growing Disaster Impacts

In November 2013, a powerful typhoon Haiyan (called Yolanda locally) swept across the Philippines—cutting a path of destruction through several central islands—killing 6,000 people, leaving 14 million affected, and causing the deadliest disaster of the year (UNOCHA 2013). Earlier in 2012, Superstorm Sandy caused economic damages of $50 billion in the USA and was the most expensive disasters among the 357 natural disasters registered globally in 2012 (Guha-Sapir et al. 2013). In April 2011, the Tohoku earthquake wreaked havoc in Japan. The following tsunami and cascading events led to the Fukushima Daiichi nuclear plant crisis, demonstrating the devastating results when Mother Nature's and man's actions do not synchronize. Even earlier in 2010, a killer earthquake created heavy destruction in Haiti, a very poor country, which went further back on its already-lagging developmental trajectory.

Such incidents are not new but quite frequent, showing the continuous catastrophic effects of disasters—deaths; unprecedented destruction to buildings, infrastructure, and livelihoods; and acute human suffering. Some of the deadly disasters that recently hit the world include an earthquake in Haiti (2010) causing more than 200,000 deaths, Cyclone Nargis in Myanmar (2008) killing 138,366 people, and the Sichuan earthquake in China (2008) killing 87,476 people. Apart from humanitarian impacts, disasters affect all sectors of a country's economy—which can take years to recover. Box 1.1 below shows how economic impacts of a disaster are assessed by international development agencies.

Even with the increased global attention to disaster prevention and preparedness in recent years, disaster impacts are rising. The frequency and effects of reported disasters are increasing globally with potentially greater fatalities and losses of people's livelihoods (UNISDR 2011). Between January 1975 and October 2008, the international emergency disasters database (EM-DAT) recorded 8,866 natural disaster events killing 2,283,767 people with the majority of related mortality and losses

Box 1.1. Assessing Impacts of a Disaster

Damage and Loss Assessment (DaLA) is a standard methodology used by international development agencies such as the United Nations and the World Bank to assess country-specific disaster impacts. The methodology measures: *damage*, which is the value of physical, durable assets that may be destroyed due to the action of the natural hazard that caused the disaster, expressed in terms of the replacement value of the assets assuming the same characteristics that they had prior to the disaster, and *losses* or changes in the normal flows of the economy that may arise in all sectors of economic and social activity due to the external shocks brought about by the disaster, until full economic recovery and reconstruction has been achieved, and are expressed in current values. Disasters affect all sectors of economy directly or indirectly but for assessment following sectors are typically evaluated: (i) productive sectors such as agriculture, tourism, commerce, and industries; (ii) infrastructure sectors such as housing, transportation, power, public buildings, education, health, sanitation and water supply, and communication; and (iii) cross-cutting sectors such as environment, livelihoods, religion, and culture. As the disasters affect almost all development sectors, even a minor disaster can cause severe economic loss depending upon the economic profile of the country.

An example of DaLA is Nigeria 2012 floods assessment. Between July and October 2012, unprecedented flooding affected 35 out of the 37 states of the country. About 363 people were killed, 5,851 injured, and 3,891,314 affected. The DaLA exercise undertaken to assess disaster's impact estimated US$16.9 billion in damages and losses due to 2012 floods. The total value of destroyed physical and durable assets was estimated at US$9.5 billion. The total value of losses across all sectors of economic activity was estimated at US$7.3 billion. The most affected individual sector in terms of destroyed assets was identified as housing, followed by agriculture.

Source: GFDRR (2014)

(relative to GDP) concentrated in low- and middle-income countries (UNISDR 2009). Between 2010 and 2012, 700 natural disasters affecting more than 450 million people were registered in international emergency disasters database. These disasters not only caused human agony and pain but large economic costs. Overall disaster damages and losses are increasing (see Fig. 1.1). Damages have risen from an estimated $20 billion on average per year in the 1990s to about $100 billion per year during 2000–2010 (IMF 2012). This trend is expected to continue due to rising concentration of people and their assets in areas more exposed to disasters and climate change. The proportion of the world's GDP annually exposed to tropical cyclones increased from 3.6 % in the 1970s to 4.3 % in the 2000s. While the economic losses from disasters tend to be higher in high-income countries due to greater exposure of valuable properties and assets, low- and middle-income countries tend to face high death tolls and disruptions to hard-earned development gains (World Bank 2010).

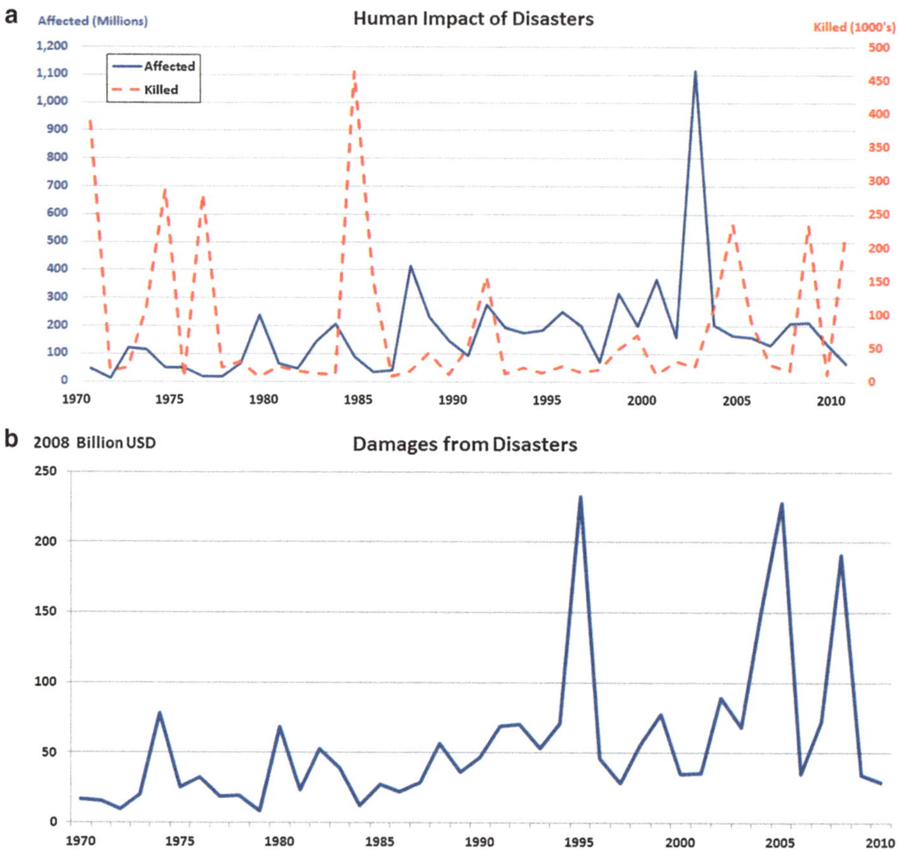

Fig. 1.1 Disaster impacts from 1970–2010. (**a**) Rising number of people affected by disasters. (**b**) Increasing global damages from disasters (Source: World Bank (2010))

With increased global attention to disaster prevention and preparedness, many developing countries are undertaking capacity building programs to reduce the extent of damages from disasters. The disaster mitigation activities include preventive actions aimed at reducing loss of lives and damages resulting from disasters. Development partners and donors now agree that it is more cost-effective to invest in disaster management and preparedness activities than to keep on providing major relief aid. Along with funding for rehabilitation of infrastructure and houses destroyed from disasters, the donors and many national governments are increasingly funding capacity building programs to increase local government's capacity to prevent and effectively respond to disasters. International and government funding

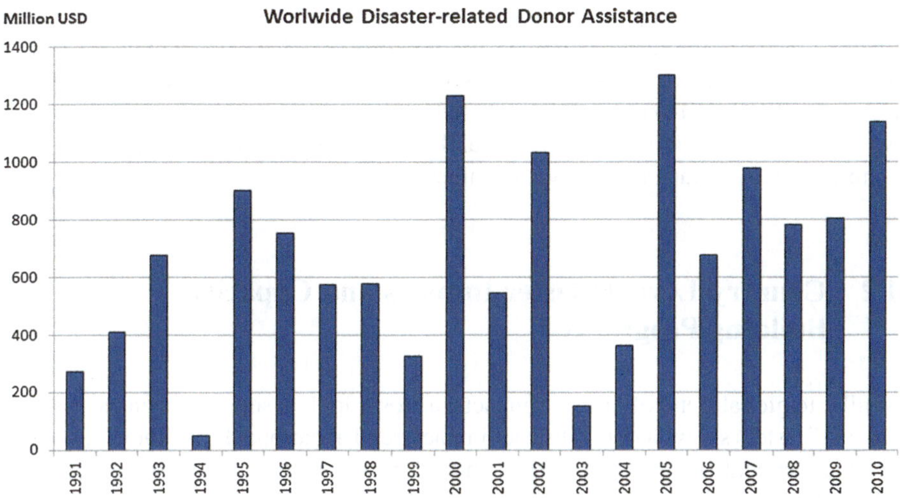

Fig. 1.2 Worldwide disaster-related donor assistance (Source: Kellet and Caravani (2013))

on disasters have increased. Figure 1.2 above shows disaster-related international financing between 1991 and 2010. Individual government's spending on disasters has also increased over the same period in some countries (see Box 1.5).

Even with increased funding available for emergency response and mitigation, disasters continue to strike causing higher damages and losses especially in areas that witnessed disasters in the past. Some examples of countries with recurrent disasters are China, Indonesia, Turkey, and Iran, where deadly earthquakes happen periodically; the Philippines, Thailand, and India, where floods occur almost every year taking lives of thousands; and many Latin American countries that suffer from floods, landslides, and earthquakes. Developed countries are thought to be better prepared to face disasters. However, although the fatalities might be less, they also face recurrent damages and losses from disasters. For example, the great Tohoku earthquake and tsunami in 2011 killed 20,000 people in Japan and induced cascading infrastructure damages and nuclear plant failure. The 2012 Superstorm Sandy caused unprecedented impacts in the USA, killing 147 people, destroying 650,000 houses, and causing $50 billion in damages (NOAA 2013).

Two trends point to further increase in global disaster impacts: first, population growth in cities and areas exposed to disasters is increasing and second, climate change. Nearly 75 % of the world's population lives in areas affected at least once by an earthquake, tropical cyclone, flood, or drought between 1980 and 2000 (UNDP 2004). This population at risk of disasters is expected to increase with population growth and urbanization, which leads to the concentration of population and assets in smaller areas. Almost all of the new population growth will happen in developing countries, which will add 2.3 billion residents to reach 7.9 billion (out of a total projected global population of 9 billion) by 2050. Second, climate change is likely to increase frequency and intensity of hydrological events including flooding and drought. Apart from sea level rise, which affects small island states and coastal cities, the IPCC (2014) report has predicted increases in extreme events such as heat waves, extreme precipitation, and coastal flooding with increase in temperature. Are countries ready to face this growing challenge?

1.2 Country-Level Disaster Impacts and Capacity Building Programs

Similar to global trend of disaster impacts, a closer look at selected countries at high risk of disasters shows a rising disaster impact trend at national level. Table 1.1 below summarizes and Boxes 1.2, 1.3, 1.4, and 1.5 provide examples of some countries which are facing recurrent disasters and are undertaking conscious steps to increase their capacity on disaster risk management. These country examples are selected on

Table 1.1 Selected country examples—disaster risk and financing on disaster risk reduction (between 1991 and 2010)

Country	Mortality risk index	Overall rank in receiving international aid on DRR (1991–2010)	International financing on DRR per capita/ national government revenue per capita (in USD)	Disaster impact trend
Low-income countries				
Ethiopia	6	n/a	0.35/19.9	Number of people affected increasing
Haiti	6	26	11.52/–8.23	Number of people affected increasing, faced deadliest disaster in 2010, still recovering
Lower-middle-income countries				
Philippines	8	4	10.78/416.42	Number of people affected increasing
Indonesia	9	2	6.75/513.64	Number of people affected increasing
India	9	8	0.5/262.28	Number of people affected increasing
Upper-middle-income countries				
Argentina	5	7	14.82/3,249.53	Number of people affected increasing
China	9	1	1.25/n/a	Number of people affected increasing

Note: National government revenue is the amount calculated by subtracting overseas development assistance (e.g. United Nations) from national revenue (tax, fees, etc.)
Source: Kellet and Caravani (2013), Guha-Sapir et al. (2014), UNISDR (2014a)

Box 1.2. Snapshot 1: Ethiopia

The memories of a little girl in a small village in the dry lands of Oromia region in Ethiopia during my visit after the 2011 drought are still vivid. She had not eaten for days and had been surviving on sugar syrup. Her mother, like many in this pastoral community, had been giving sugar dissolved in water to her kids in place of milk and other nutritional food. As the community was trying to explain their sufferings and suggestions to improve their situation, the girl's big eyes stay fixated on me. Her only question to me—through a translator—was if the rains will ever come back. Later, the community showed me a small primary school that has not been functioning well due to the lack of willing teachers to serve this remote community. Now the community wanted the government to invest in better schooling so that the new generation need not rely solely on pastoralism.

This was not the first drought in Ethiopia. This low-income country of 94 million inhabitants (World Bank 2013) has historically suffered from droughts. The country suffers from chronic and transitory food insecurity, requiring support for an average of 8.3 million people yearly (Siyoum et al. 2012). Why is this situation occurring again and again even when Ethiopia has been making steady economic growth in the recent years? While there were humanitarian responses every year, the government and donors took a more systematic approach from 2005 with the launch of a large donor-supported program on safety nets. Why has the number of people requiring support not gone down after this targeted program, especially when the government started investing more resources strategically on drought response and risk reduction every year?

Box 1.3. Snapshot 2: Haiti

Nearly 150,000 people were still living in tents and waiting to move back into their houses even 4 years after the deadly 2010 earthquake in Haiti (NPR 2014). The country saw tremendous international support after the deadly earthquake which killed 220,000 people, affected 120 % of the country's economy, and led to very long and arduous post-earthquake recovery and reconstruction efforts (GFDRR 2010a). International agencies and donors have invested in capacity and resilience building programs such as support for multi-sector coordination and technical training programs and safer schools and buildings.

Nearly 96 % of Haiti's population is still living in areas at high risk from cyclones and earthquakes (GFDRR 2010b). The 2012 cyclone affected 200,000 people while smaller flooding events were reported in 2010. In 2004, tropical storm Jeanne affected over 315,000 people and in 2008, another tropical storm affected more than 865,000 people (GFDRR 2010b). The country's high population density (up to 40,000 per sq km in Port-au-Prince), large number of informal structure activities, weak infrastructure, and environmental degradation render the country and its population particularly vulnerable.

Is there a way for Haiti to break its capability trap?

Box 1.4. Snapshot 3: Philippines

The Typhoon Yolanda/Haiyan, which affected many regions of the Philippines in 2013, was not the first typhoon to affect the country. The Philippines' tropical climate and its location in the "ring of fire" make it susceptible to many typhoons every year. As per the United Nations, the country is ranked twelfth globally in terms of disaster mortality index and is ranked seventh in receiving global donor support to mitigate disaster risk. In the last 20 years, it received $1.5 billion from donors on disaster risk management, 55 % of which went specifically for flood risk reduction. The country has received fourth highest donor financing for disaster risk reduction worldwide over the last 20 years (Kellet and Caravani 2013).

While the funding to mitigate disaster risk has grown, the impacts from typhoons have also grown over the years. The 2009 Typhoons Ketsana/Ondoy and Pepeng affected 13 million people with 956 dead and 220,000 houses damaged. The 2013 Typhoon Yolanda/Haiyan affected 16 million people with 6,000 dead and one million houses damaged (USAID 2014). Post-Disaster Needs Assessments conducted by international agencies and partners with the government after 2009 typhoons estimated $4.4 billion for recovery and reconstruction (GFDRR 2009). The humanitarian needs after 2013 typhoon are estimated at $788 million and recovery and reconstruction needs more than $1 billion (USAID 2014).

Why is the country not able to reduce mortality rates from recurrent typhoons even with conscious efforts?

Box 1.5. Snapshot 4: Indonesia

Indonesia is the world's fourth most populous country and ranks 12th globally for high mortality risks from multiple hazards. More than 90 million people (40 %) in the country are at risk of tsunamis, earthquakes, volcanic eruptions, and floods. No wonder, the country received second highest international aid for disaster risk management over the last 20 years. Moreover, the government of Indonesia has invested around 10 times more on disaster risk reduction on average compared to the international aid ($900 million between 2006 and 2012) (Kellet and Caravani 2013). However, even as the national and international funding on DRM increased, the impacts of disasters also have grown over the years. The graph below shows the impact of disasters from 1994 to 2009.

(continued)

Box 1.5. (continued)

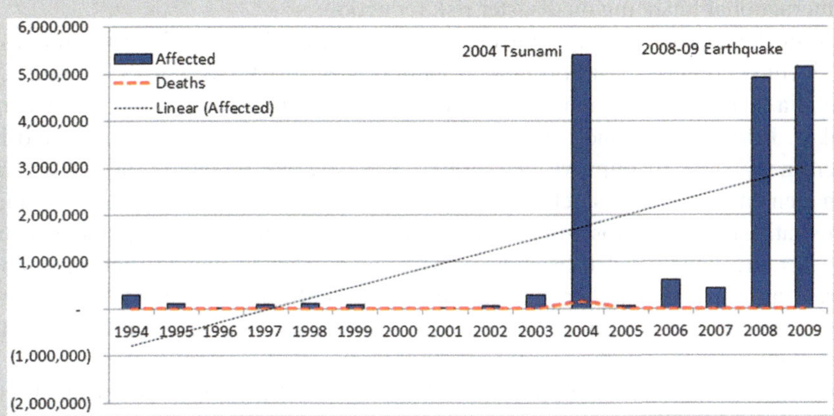

Source: UNISDR (2014b)

West Sumatra province of Indonesia is at a high risk of earthquakes due to its location at the convergence zone of four major tectonic plates and densely populated settlements. Even though the region has a long history of high intensity earthquakes, even a moderate intensity earthquake causes relatively high impacts. The 2007 earthquake had a relatively less magnitude of 6.3 but caused much higher impacts in West Sumatra (Padang) region—66 deaths, 500 casualties, severe damage or collapse of nearly 15,000 buildings, and over 135,000 people displaced (GFDRR 2010c). In 2009, another powerful (magnitude 7.6) earthquake struck West Sumatra province. The earthquake affected 13 out of 19 districts, killing over 1,100 people and injuring another 3,000 (BNPB et al. 2009). The death toll was intensified by landslides in at least three villages in the district of Padang Pariaman, burying a significant proportion of the inhabitants. The cities of Padang and Pariaman (Kota Padang and Kota Pariaman) as well as the district of Padang Pariaman (Kabupaten Padang Pariaman) were the worst-affected areas.

A recent assessment on DRM capacity in the country pointed to the usual impediments—lack of capacity and capability on DRR of nongovernment stakeholders at local levels, lack of systematic learning, and inability to mainstream DRM into development agenda (Djalante et al. 2012). Why has it been hard to develop DRM capacity in the country even after the conscious efforts of government and donors?

the basis of their risk to disasters (based on UNISDR's mortality risk index), their economic well-being (based on World Bank's income classification), and their access to international financing on disaster risk reduction.

At the lower end of the income classification, Ethiopia and Haiti are examples of low-income countries with prominent natural hazard risk (see Boxes 1.2 and 1.3). Ethiopia faces recurrent and long-lasting droughts, while Haiti faces deadly earthquakes. Ethiopia has made steady economic progress in the last decade and has taken many steps to improve institutions, policies, and financing on disaster risk management. Its social safety net program is a successful example of providing systematic support to most vulnerable people. The disaster risk management efforts, however, are mostly aided by donors, with the number of people needing support increasing every year rather than falling down. Haiti, on other hand, received tremendous international support after 2010 earthquake, but even after 4 years it has not totally recovered with many people still living in temporary shelters.

Indonesia and the Philippines, both lower-middle-income countries highly exposed to multiple hazards, received second and fourth highest support from donors on disaster risk management in the last 20 years. The countries have invested relatively higher domestic resources, compared to international aid available to them, to develop their own capacity to prevent and prepare for disasters. Even with more funding and focus on disasters, the countries are still facing growing disaster impacts (see Boxes 1.4 and 1.5).

Though Argentina, a higher-income country, has invested heavily in disaster risk reduction, it faces growing disaster impacts (see Box 1.6). At the higher end of the income classification, the USA and Japan are examples of developed countries. Both face recurrent and deadly disasters and invest heavily in disaster risk management. In both countries, fatalities are lower over the years but losses are very high and growing (Weiss and Weidman 2013).

One common theme among the entire spectrum of example countries is the rise in losses as well as people affected from disasters. It also shows that countries are investing in increasing their capacity, but are these countries investing in the "right" capacity building measures? What is meant by effective capacity for disaster risk management? How should the capacity be assessed since disaster trend alone may not reflect it?

Recent research by UNDP (2013) points to risk governance capacity, which is defined as the way in which the public authorities, civil servants, media, private sector, and civil society coordinate at community, national, and regional levels in order to manage and reduce disaster- and climate-related risks, as having a direct relationship with the development stage of the country (UNISDR 2011). Not surprisingly some of the lowest-income countries end up being the ones with the least capacity. See Fig. 1.3 for details. The figure also shows that some high- and medium-income countries also have low disaster risk capacity. What factors lead countries from low disaster mitigation capacity to high capacity? Currently these factors are not elaborated in the literature and so knowledge about them remains limited.

Box 1.6. Snapshot 5: Argentina

Argentina is among the top ten emerging economies with the highest flood hazard exposure. Swiss Re estimates the potential of losses from floods in excess of US$ 3 billion a year (World Bank 2014). The disaster impacts, not only in terms of economic losses but also the number of people affected, are rising. Climate change can further increase the trend. On April 2, 2013, the city of Buenos Aires experienced one of the heaviest storms recorded in nearly 50 years. Nearly 350,000 people were directly affected. Damage and losses added up to nearly US$ 300 million. Key transportation routes were submerged and mass-transit systems like the Buenos Aires metro and railway systems were disabled. Power shortages lasted for as long as 15 h in at least 11 neighborhoods.

The country has been very proactive in investing in disaster risk management. It is one of the top 10 countries receiving aid on flood risk prevention in the last two decades (Kellet and Caravani 2013). The country started building its capacity for preventing and mitigating disaster impacts from 1998, with the establishment of the Federal Emergency System (Freeman et al. 2003). The country has developed a decentralized system where the federal government plays an important role in the mobilization of resources and the coordination of national and international organizations in mitigation and emergency response efforts (World Bank 2012). Provincial governments assume the responsibility for assessing regional vulnerabilities and carrying out mitigation projects to protect against natural catastrophes within their respective provinces. From 1998, a number of World Bank loans have gone toward developing capacity for flood resilience. However, many challenges related to coordination between government organizations, availability of timely and useful data, and staff capacity hamper effective capacity building.

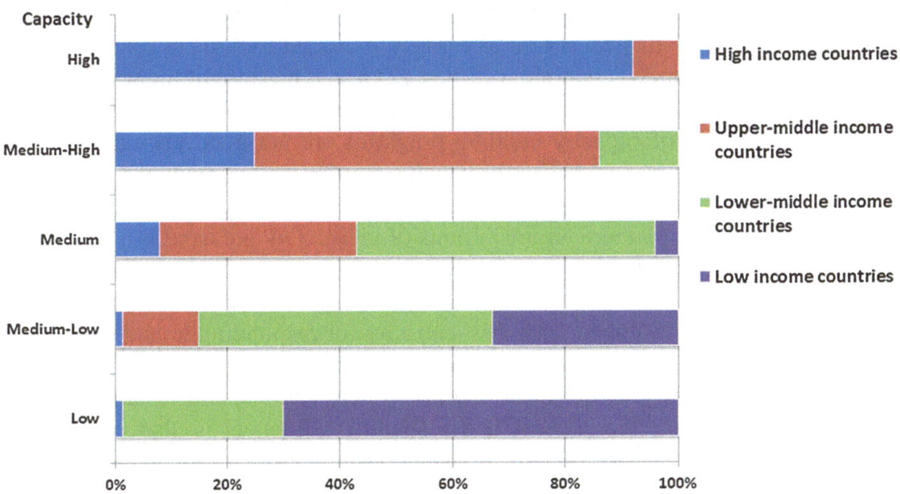

Fig. 1.3 Disaster risk governance capacity (Source: UNISDR (2011))

1.3 Capacity Development Mirage

How can a place be built and managed so that it is safe for people to live? Ironically, many governments and people keep on asking the same question after every new disaster. Why, even with a high level of investment in increasing government's capacity to manage disaster risk, do the impacts of disasters continue to increase? What is the challenge?

Some countries seem to be falling in a "capability trap"—where in spite of conscious capacity development programs, disaster preparedness and resilience remain low. The concept of capability trap is based on the findings from recent studies which suggest that some capacity building programs can in fact reverse the small progress made in developing countries (Abraham and Platteau 2004; Rao and Ibanez 2005; Pritchett and Weijer 2010). A capability trap occurs when the capability of the state to implement policies and programs is both severely limited and improving only very slowly (Birdsall 2007; Moyo 2009; Pritchett and Weijer 2010). This concept is discussed in more detail in Chap. 3.

In this regard, a recent remark by a senior project manager of an international development agency, during a workshop on disaster early warning and response, was very compelling. Summarizing the current state of affairs, the manager said that even after 7 years of funding capacity building programs for strengthening early warning and disaster preparedness capacity in many Central Asian countries, there is no improvement in providing timely warnings to citizens and triggering timely government response. The manager is not the first to show frustration over capacity building programs for disaster prevention. Commenting on the unsatisfactory response to Hurricane Katrina in the USA, during which nearly 1,500 people died and thousands lost homes, the Committee on Homeland Security and Governmental Affairs (2006) noted that failure to prevent losses and respond in time was due to the "failure of government at all levels to plan, prepare for, and respond aggressively to the storm" (p. 2).

Many questions remained to be answered: How to define and measure capability trap? Why are some countries facing capability trap? Is it because disasters are unique and pose a challenge that cannot be effectively managed by human societies? Is it dependent on a country's economic well-being? Or is it because of the way capacity development for disaster risk management is taking place—in other words the design and implementation of capacity building programs are not good enough? Is there altogether different capacity needed for which countries are not ready yet? What can be done to identify and break the capability trap? What causes capability trap?

On the last question, there are two points of views. On one hand is a "Yes Aid" group. Comprising predominantly of donors and international aid agencies, proponents of this group work on the premise that international aid is necessary to increase developing or least developed countries' capacity for responding to and preventing disaster impacts. This group believes that the required financial resources are not available in the countries to take needed and timely actions, and technical understanding of what to do to prevent and respond to disasters is also not known. International aid can provide both the financial resources and the technical expertise to resolve the problem.

On the other hand is the "No Aid" group. This group consists of recent researchers focusing on the review of development aid over the last several decades and advocates that the international aid has not been helpful to recipient countries and has in fact reversed the development gains (Birdsall 2007; Moyo 2009; Pritchett and Weijer 2010; Easterly 2013). Proponents of this group believe that the way capacity building programs are designed or implemented contribute to the capability trap situation. Simply put, it is the situation where capacity development projects start with a solid promise of development and amidst high expectations fail to provide desired results on the ground. As in a mirage, the international development assistance or even government's own actions rely on best practices and supply-driven and short time-frame projects focusing on fiscal disbursement, numerical targets, and short fixes, leading in the opposite direction (Pritchett and Weijer 2010).

Compared to typical development projects such as roads and agriculture, the need for international aid for disaster is very different. The humanitarian requirements after major or catastrophic events require donors to provide post-disaster response assistance. No wonder, a majority of international disaster financing (see Fig. 1.2) goes toward emergency response and recovery, which keeps growing with higher disaster events and impacts worldwide. Since the disaster response costs are increasing, the donors naturally want to assist countries in developing long-term capacity to mitigate disaster impacts, so over time less funding is required on disaster response.

Moving from global to country level, dynamics around a disaster are also very different and unique. Disasters are unique and "focusing events" (Birkland 2007), which get a lot of attention from media. Governments come under high pressure to perform quickly—recover from disasters, rehabilitate affected people, and take urgent steps to avoid future impacts. This requires immediate finance and technical know-how, which may not readily exist in the country. International assistance, in such cases, remains the only option. There may also be public pressure to punish or take blame for disaster impacts (AusSMC 2012). Such a situation can very well lead into quick but unsustainable policy changes and projects to show quick results. Funding for disaster risk reduction suffers from constant diversion to emergency response (especially if an emergency response department is responsible for risk reduction activities).

Another challenge is related to each disaster's "half-life." In spite of the high media coverage of disaster events, the memory of a disaster soon fades away. Once the immediate relief and reconstruction efforts are over, urgent developmental issues such as access to water, electricity, and housing (in the development country context) take central stage, consuming most of the time and resources of communities. "Even a major disaster has a half-life of memory of less than two generations and other more immediate threats often seem more urgent. Less severe events can be forgotten in less than 3 years" (Jha et al. 2012, p. 49). How then can institutions and organizations be on top of disasters, which are uncertain and less frequent, as compared to urgent development issues that are certain, more visible, and frequent? Can all these reasons be a part of "capability trap" situation? How can this be verified?

There are many reasons why countries need to be prepared for disasters. Not only do the disasters affect countries in the medium to long run, they also increase poverty, affecting the most vulnerable. Some low-income countries have little capacity to manage disasters and are severely affected by a disaster, which further reduces their ability to prevent future disasters and achieve overall development. Haiti is one example. What can be done to break the capability trap?

1.4 Breaking the Capacity Development Mirage

This chapter provided an overview of how disaster impacts are increasing globally and how some countries are not able to develop effective capacity for preventing and mitigating disaster impacts. With examples of selected countries facing recurrent disasters, this chapter raises critical questions on challenges associated with capacity development in disaster risk management: How to define capability trap? Why are some countries facing capability trap? What can be done to develop effective capacity and break the capability trap? To better understand the challenges at a local level, the next chapter introduces the case of 2001 earthquake in Gujarat, India, where both donors and the government invested heavily in capacity building programs. After an overview of the earthquake-affected region, earthquake impacts, and the kind of capacity building measures undertaken, the chapter focuses on a critical question: after massive donor and government investment for over a decade, what disaster risk management capacity exists at the local level—the front line charged with facing future disasters?

The second part of the book focuses on finding the answers to questions raised in Chaps. 1 and 2. The solutions are broadly based on two lines of inquiries: first, by developing a conceptual model of capacity building in disaster risk management and, second, by applying the conceptual model to understand capacity development challenges in Gujarat. Implications for Gujarat and generally for donors, governments, and communities are then discussed based on the results of the above inquiries. A comparison of countries with better disaster prevention and response capacities is also undertaken to support recommendations.

The third and fourth chapters dig deep into the meanings and practices of capacity building and disaster risk management. Fifth chapter provides a critical rethinking of capacity building model as it applies to disaster risk management. Sixth chapter focuses on understanding Gujarat case through the conceptual model. The seventh and eighth chapters focus on recommendations for improving capacity building in disaster risk management.

There has been limited research on enhancing local capacity to manage disasters. In particular, the concept of capacity development for disaster risk management has not been adequately explored in public policy literature. This book contributes to this emerging field of public policy. Similarly, there has been research about local government capacity related to service delivery and financial management; the role of local government in managing disaster risk is not a well-researched field. This book contributes to the literature on local government capacity literature. The research and recommendations in this book will also aid in understanding why certain local capacity factors are critical in certain contexts and can guide future generations of development practitioners in prioritizing actions to achieve anticipated results.

References

Abraham, A., & Platteau, J.-P. (2004). Participatory development: Where culture creeps in. In V. Rao & M. Walton (Eds.), *Culture and public action*. Stanford: Stanford University Press.

AusSMC. (2012). *Rapid reaction: Earthquake scientists guilty of manslaughter in Italy – Experts respond*. Australian Science Media Center. http://www.smc.org.au/2012/10/rapid-reaction-earthquake-scientists-guilty-of-manslaughter-in-italy-experts-respond/. Accessed on 22 Apr 2013.

Birdsall, N. (2007). Do no harm: Aid, weak institutions and the missing middle in Africa. *Development Policy Review, 25*(5), 575–598.

Birkland, T. (2007). *Lessons of disaster: Policy change after catastrophic events*. Washington, DC: Georgetown University Press.

BNPB, Bappenas, Provincial and State Governments of West Sumatra and Jambi and International Partners. (2009). *West Sumatra and Jambi natural disasters: Damage, losses, preliminary needs assessment*. https://gfdrr.org/indonesia%E2%80%93-2009%E2%80%93-pdna-undertaken-after-earthquake-killed-1100-west-sumatra. Accessed Sept 2014.

Committee on Homeland Security and Governmental Affairs. (2006). *Hurricane Katrina: A nation still unprepared*. Washington, DC: U.S. Government Printing Office.

Djalante, R., Thomalla, F., Sinapoy, M. S., & Carnegie, M. (2012). Building resilience to natural hazards in Indonesia: Progress and challenges in implementing the Hyogo Framework for Action. *Natural Hazards, 62*, 779–803.

Easterly, W. (2013). *The tyranny of experts: Economists, dictators, and the forgotten rights of the poor*. New York: Basic Books.

Freeman, P. K., Martin, L. A., Linnerooth-Bayer, J., Mechler, R., Pflug, G., & Warner, K. (2003). *Disaster risk management: National systems for the comprehensive management of disaster risk and financial strategies for natural disaster reconstruction*. Washington, DC: Inter-American Development Bank. http://idbdocs.iadb.org/wsdocs/getdocument.aspx?docnum=36344547

GFDRR. (2009). *Philippines 2009 – Typhoons Ondoy and Pepeng affected 9.3 million people*. https://www.gfdrr.org/philippines%E2%80%93-2009%E2%80%93-typhoons-ondoy-and-pepeng-affected-93-million-people. Accessed June 2014.

GFDRR. (2010a). *Haiti 2010 PDNA – At a glance*. https://www.gfdrr.org/sites/gfdrr.org/files/documents/Haiti_2010_PDNA_GLANCE.pdf. Accessed Sept 2014.

GFDRR. (2010b). *Disaster risk management in Latin America and the Caribbean region: GFDRR Country notes Haiti*. http://www.gfdrr.org/sites/gfdrr.org/files/Haiti-2010.pdf. Accessed Sept 2014.

GFDRR. (2010c). *It is not too late: Preparing for Asia's next big earthquake with emphasis on Philippines, Indonesia and China*. http://www.gfdrr.org/sites/gfdrr/files/publication/Preparing_for_Asia%27s_Next_Big_Earthquake_0.pdf. Accessed Sept 2014.

GFDRR. (2014). *Nigeria post-disaster needs assessment: 2012 floods*. https://www.gfdrr.org/sites/gfdrr/files/NIGERIA_PDNA_PRINT_05_29_2013_WEB.pdf. Accessed May 2014.

Guha-Sapir, D., Hoyois, P., & Below, R. (2013). *Annual disaster statistical review 2012: The numbers and trend*. Brussels: CRED. http://cred.be/download/download.php?file=sites/default/files/ADSR_2012.pdf

Guha-Sapir, D., Below, R., & Hoyois, P. (2014). *EM-DAT: International disaster database*. Brussels: CRED. www.emdat.be. Accessed Mar 2014.

IMF. (2012). *Natural disasters: Mitigating impact, managing risks* (IMF Working Paper, WP/12/245). https://www.imf.org/external/pubs/ft/wp/2012/wp12245.pdf. Accessed Sept 2014.

IPCC. (2014). Summary for policymakers. In C. B. Field, V. R. Barros, D. J. Dokken, K. J. Mach, M. D. Mastrandrea, T. E. Bilir, M. Chatterjee, K. L. Ebi, Y. O. Estrada, R. C. Genova, B. Girma, E. S. Kissel, A. N. Levy, S. MacCracken, P. R. Mastrandrea, & L. L. White (Eds.), *Climate change 2014: Impacts, adaptation, and vulnerability* (Part A: Global and sectoral aspects.

Contribution of working group II to the fifth assessment report of the intergovernmental panel on climate change, pp. 1–32). Cambridge/New York: Cambridge University Press.

Jha, A., Bloch, R., & Lamond, J. (2012). *Cities and flooding: A guide to integrated urban flood risk management for the 21st century*. Washington, DC: The World Bank.

Kellet, J., & Caravani, A. (2013). *Financing disaster risk reduction: A 20 year story of international aid*. Washington, DC: Global Facility for Disaster Reduction and Recovery (GFDRR) and Overseas Development Institute (ODI).

Moyo, D. (2009). *Dead aid: Why aid is not working and how there is a better way for Africa*. New York: Farrar, Straus and Giroux.

NOAA. (2013). *Hurricane/post-tropical cyclone Sandy, October 22–29, 2012*. Silver Spring: NOAA.

NPR. (2014, January 12). *Four years after earthquake, many in Haiti remain displaced*. http://www.npr.org/2014/01/12/261723409/four-years-after-earthquake-many-in-haiti-remain-displaced. Accessed Sept 2014.

Pritchett, L., & Weijer, F. (2010). *Fragile states: Stuck in a capability trap?* Input paper for The World Development Report 2011. Washington, DC: The World Bank.

Rao, V., & Ibanez, A. M. (2005). The social impact of social funds in Jamaica: A 'participatory econometric' analysis of targeting, collective action, and participation in community-driven development. *Journal of Development Studies, 41*(5), 788–838.

Siyoum, A. D., Hilhorst, D., & Uffelen, G. (2012). Food aid and dependency syndrome in Ethiopia: Local perceptions. *Journal of Humanitarian Assistance*. http://sites.tufts.edu/jha/archives/1754. Accessed Sept 2014.

UNDP. (2004). *A global report on reducing disaster risk: A challenge for development*. New York: United Nations Development Program.

UNDP. (2013). *Issue brief: Disaster risk governance*. Bureau for Crisis Prevention and Recovery. New York: UNDP.

UNISDR. (2009). *Global assessment report on disaster risk reduction*. Geneva: United Nations.

UNISDR. (2011). *Global assessment report on disaster risk reduction*. Geneva: United Nations.

UNISDR. (2014a). *Mortality risk index*. http://www.preventionweb.net/files/9929_MRIA3.pdf. Accessed Sept 2014.

UNISDR. (2014b). *Inventar Disaster Information System*. http://www.desinventar.net/DesInventar/profiletab.jsp?countrycode=id. Accessed on 3 Mar 2014.

UNOCHA. (2013). *Asia Pacific humanitarian bulletin 2013*. http://reliefweb.int/sites/reliefweb.int/files/resources/HB-AP-2013.FINAL__0.pdf. Accessed on 5 Apr 2014.

USAID. (2014). *Philippines – Typhoon Yolanda/Haiyan Fact Sheet #20*. http://www.usaid.gov/sites/default/files/documents/1866/01.24.14_USAID-DCHATyphoonYolandaHaiyan_FactSheet20.pdf. Accessed Sept 2014.

Weiss, D. J., & Weidman, J. (2013). *Disastrous spending: Federal disaster-relief expenditures rise amid more extreme weather*. Center for American Progress. http://www.americanprogress.org/issues/green/report/2013/04/29/61633/disastrous-spending-federal-disaster-relief-expenditures-rise-amid-more-extreme-weather/. Accessed on 6 Apr 2014.

World Bank. (2010). *Natural hazards, unnatural disasters: The economics of effective prevention*. Washington, DC: The World Bank.

World Bank. (2012). *Improving the assessment of disaster risk to strengthen financial resilience. A special G20 publication by the Government of Mexico and World Bank*. http://www.gfdrr.org/sites/gfdrr.org/files/Chapter_2-Argentina-Assessment_of_Disaster_Risk_Management_Strategies_in_Argentina.pdf. Accessed Sept 2014.

World Bank. (2013). *Ethiopia: Country at a glance*. http://www.worldbank.org/en/country/ethiopia. Accessed Sept 2014.

World Bank. (2014). *Project information document – AR Vega flood prevention and drainage project*. http://www-wds.worldbank.org/external/default/WDSContentServer/WDSP/LCR/2014/01/29/090224b08222c990/1_0/Rendered/PDF/Project0Inform0ge0Project000P145686.pdf. Accessed May 2014.

Chapter 2
Capacity Crisis After the 2001 Earthquake in Gujarat, India

Abstract The state of Gujarat in India faced a deadly earthquake in 2001, which killed over 13 thousand people, affected nearly 28 million, and damaged 1.2 million houses. In the worst-affected region of the state, Kutch district, over 90 % population was affected and 70 % buildings destroyed. While the world has witnessed even worst impacts from deadly earthquakes before, what makes this disaster different is the adoption of a massive government-led and donor-supported reconstruction and capacity building program after the earthquake. Although the country and the state were not at all prepared for such a disaster in 2001, enormous international and national attention after the earthquake led to fundamental changes in how disasters are managed across the country. The government of Gujarat acted swiftly in establishing a new state disaster management agency within a month. New state and national disaster management laws were passed in 2003 and 2005, respectively, defining federal, state, and district level institutional arrangements for disaster management. Most importantly, these laws provided dedicated funding not only for disaster response but also disaster risk mitigation. This chapter starts with an overview of the need to focus on Gujarat case along with an overview of disaster trends in India, Gujarat state, and Kutch district. A brief description of the 2001 earthquake in Gujarat is followed by capacity building efforts that were undertaken by the government and donors. Towards the end of the chapter, a crucial question is raised. More than a decade later, and with $1.7 billion spent in targeted capacity building program, is the region any safer?

Keywords 2001 Earthquake of Gujarat • Disaster management • Donors • Gujarat Emergency Earthquake Reconstruction Project (GEERP) • Gujarat State Disaster Management Authority (GSDMA)

The state of Gujarat in India faced a deadly earthquake in 2001, killing over 13 thousand people, affecting 28 million, and damaging 1.2 million houses in a matter of minutes (Mishra 2004). In the worst-affected region of the state, Kutch district, over 90 % population was affected and 70 % buildings destroyed. While the world has witnessed even worst impacts from deadly earthquakes before, what makes this disaster different is the adoption of a massive government-led and donor-supported reconstruction and capacity building program after the earthquake. Although the country and the state was not at all prepared for such a disaster in 2001, enormous

A. Tiwari, *The Capacity Crisis in Disaster Risk Management*,
Environmental Hazards, DOI 10.1007/978-3-319-09405-2_2

international and national attention after the earthquake led to fundamental changes in how disasters are managed across the country. The government of Gujarat acted swiftly in establishing a new state disaster management agency within a month. New state and national disaster management laws were passed in 2003 and 2005, respectively, defining federal, state, and district level institutional arrangements for disaster management and most importantly providing dedicated funding not only for disaster response but also disaster risk mitigation.

This chapter starts with an overview of the need to focus on Gujarat case, along with an overview of disaster trends in India, Gujarat, and Kutch. A brief description of the 2001 earthquake in Gujarat is followed by capacity building efforts that were undertaken by the government and donors. Toward the end of the chapter, a crucial question is raised. More than a decade later, and with $1.7 billion spent in targeted capacity building program, is the region any safer? How to assess whether the capacity building efforts were effective, especially in the absence of another earthquake? Should the people wait till the next earthquake to find out? A potential capability trap position is discussed in which local capacity for preventing and preparing for earthquake is either developing very slowly or is not being sustained effectively.

2.1 Why Focus on the 2001 Earthquake of Gujarat?

Many deadly earthquakes have occurred around the world after the 2001 earthquake of Gujarat. Why then focus on it? First, it happened in one of the most populous and disaster-prone countries of the world. India is highly prone to disasters, and with a growing population, even more people will be at risk in the future. Second, the quake happened in one of the most progressive states of the country whose economic and social development capacity can be considered relatively higher compared to other Indian states. Thus, the likelihood of success in capacity building program is higher. Third, it attracted massive international and national aid specifically for developing long-term capacity for disaster risk management. Compared to other cases where the focus is usually just on post-disaster reconstruction, this case focused specifically on long-term capacity building on disaster risk management. Fourth, and most importantly, more than a decade has passed after the capacity building program was adopted—which provides ample time for the results.

2.1.1 India: Increasing Population and Disaster Impacts

India is the second most populous country and one of the most disaster-prone countries in the world. The country with 1.2 billion population (as per 2011 census) ranks 9th on the Disaster Mortality Risk Index (UNISDR 2009)—second only to China in terms of number of disaster victims (NIDM 2009)—and has seen disaster losses and number of affected people increasing over the years (see Table 2.1). India

Table 2.1 Disaster impacts in India

Period	Number of events	Total people affected	Damaged buildings
1965–1975	45	366,886,115	1,883,989
1976–1985	103	308,360,528	4,929,511
1986–1995	107	564,157,326	14,138,645
1996–2005	173	663,54,8072	23,594,614
2006–2013	121	107,122,392	13,446,247
Total	**549**	**2,010,074,433**	**57,993,006**

Source: Guha-Sapir et al. (2014)

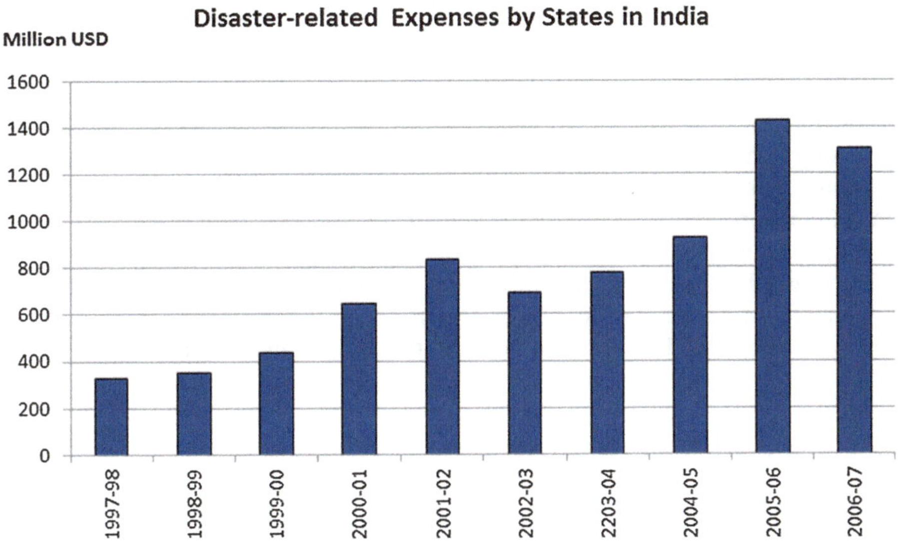

Fig. 2.1 Expenses incurred by states in India on natural disasters (Source: Data from NIDM (2009))

ranks 8th overall in international aid receipts over the last 20 years. Although the international aid is very small per capita ($0.5), it is still very substantial given that it has been concentrated in specific areas after deadly disasters (e.g., 2001 Gujarat earthquake and Indian Ocean tsunami).

The country mostly followed an emergency response approach to disasters until the 2001 Gujarat earthquake. The federal government provides financial support while affected states manage relief and reconstruction works (World Bank 2009). However, over the years, state expenditure related to disaster response has grown outpacing planned budgets for emergencies under successive finance commissions—which recommends budgetary allocations from federal to state governments over a period of 5 years. Figure 2.1 above shows the growing expenditure on disaster response by states in India. From 1997 to 2007 alone, the states in India spent nearly US $8 billion[1] on disaster response (NIDM 2009). With 55 % of its area exposed to earthquakes, 8 % to cyclones, and 5 % to floods and assuming even

[1] At a conversion rate of 1 USD to 60 Indian Rupees.

moderate impacts from climate change, the disaster losses are likely to grow even further in the future.

2.1.2 Gujarat: One of India's Most Progressive States

The state of Gujarat is situated in the western part of India and is the birth place of Mahatma Gandhi. Although it is a relatively small state accounting for 6 % of the India's area and 5 % of India's population, it is one of the most progressive states in the country. The population of Gujarat as per Census 2011 is 60 million of which 43 % live in 500 urban towns and cities and the remaining 57 % in over 18,000 villages. The per capita income of Gujarat is the fourth highest among the major Indian states at 1993–1994 prices. It is the second most industrialized state in the country accounting for over 10 % of working factories, 9 % of average daily employment, 14 % of the value of output, and 11 % of net value added of manufacturing sector in the country as a whole. It is also the third most urbanized state of the country and one of the most literate states of India with the literacy rate of 79 %.

Gujarat is highly prone to cyclones, drought, earthquakes, and floods. It has faced many disasters historically (see Table 2.2). Frequent disasters have had negative impact on the state's economy. However, for a long time systematic understanding and management of disasters didn't exist in the country and the state.

2.1.3 Kutch: Historically Prone to Earthquakes

Kutch is one of the 26 districts in the state of Gujarat. It is the largest district in India in terms of land area (45,652 km^2) (Census of India 2011). It borders Pakistan in the north and northwest, the Arabian Sea in the west, and the Gulf of Kutch in the south. The Rann of Kutch separates the district from the mainland with the Great Rann in

Table 2.2 Major disasters in Gujarat

Disaster	Years	Comments
Cyclone	1817, 1850, 1881, 1893, 1896, 1897, 1903, 1920, 1933, 1947, 1948, 1961, 1964, 1975, 1976, 1978, 1981, 1982, 1983, 1990, 1993, 1996, 1998, 1999	Very frequent hazard that is likely to increase with climate change
Drought/heat wave	1985, 1986, 1987, 1998, 1999, 2000, 2001, 2002	Good rains from 2002 to 2012
Earthquake (magnitude >5)	1668, 1819, 1821, 1845, 1856, 1864, 1864, 1903, 1927, 1940, 1956, 1970, 1982,1991, 1995, 1996, 2001	The state lies in zone V, with return period 20 years
Floods	1980, 1989, 1991, 1993, 1994, 1996, 1997, 1998, 2003, 2004	Floods have become more frequent with increase in the number of urban settlements

Source: Compiled from various GSDMA publications

the north and the Little Rann in the east and southeast. The Rann is a shallow salt marshland that is submerged in water during the rainy season and becomes dry (salt flats) during other seasons. There are a number of islands in the Great Rann. The Kutch peninsula is generally dry with an annual average rainfall of 35 cm. It has a linear hill range running east–west with a number of small seasonal streams following the slopes to north and south. The northern edge of the landmass bordering with the Great Rann is a large swath of grassland known as Banni.

Kutch has a population of nearly two million as per 2011 Census (see Table 6.3). However, compared to national and state averages, the district has a very low density. Majority of the population live in rural areas and depend on seasonal farming and animal husbandry.

Kutch district is prone to earthquakes, cyclones, floods, and drought. The district is an active seismological area with a number of active faults. Kutch has witnessed many destructive earthquakes in the past, and the 2001 earthquake (magnitude of 6.9 on Richter scale) resulted in unprecedented deaths and destruction. The area has experienced aftershocks with such regularity that the people are now capable of understanding the differences in intensity.

2.1.4 The 2001 Earthquake in Gujarat

Every year, the Republic Day in India is celebrated on January 26 to mark the day when India's constitution came into effect. Morning is especially busy on this day with celebrations and parades in schools and government offices. In 2001, on Republic Day, a major earthquake of magnitude 6.9 on Richter scale occurred in the state of Gujarat in India at 8:46 am local time (Fig. 2.2). It lasted for 2 min. The epicenter was about 9 km south-southwest of the village of Chobari, Bhachau Taluka, of Kutch district.

More than 7,600 villages of 19 districts were partially or fully affected; 13,805 human lives were lost in the state and approximately 167,000 people suffered minor or severe injury. There was significant damage to the infrastructure with facilities such as hospitals, schools, the electric power grid, water systems, bridges, and roads damaged or destroyed. Over 1.2 million houses were damaged to varying degrees, and more than 200,000 of them collapsed completely. The related consequence of the phenomenon was the loss of livelihood of millions of people. More than 10,000 small and medium industrial units stopped production and livelihoods of more than 50,000 artisans were adversely affected (GSDMA 2001).

The immensity of destruction, human suffering, and media attention prompted a quick response within India. The national and state governments quickly provided assistance in many forms including cash, medical supplies, communication teams, shelters, food, clothing, transport, and relief workers. There were more than 185 nongovernment organizations (NGOs), mostly Indian charities, which undertook earthquake-relief and rehabilitation activities. Search and rescue teams soon arrived from several countries to help local rescue teams. Relief teams and supplies soon

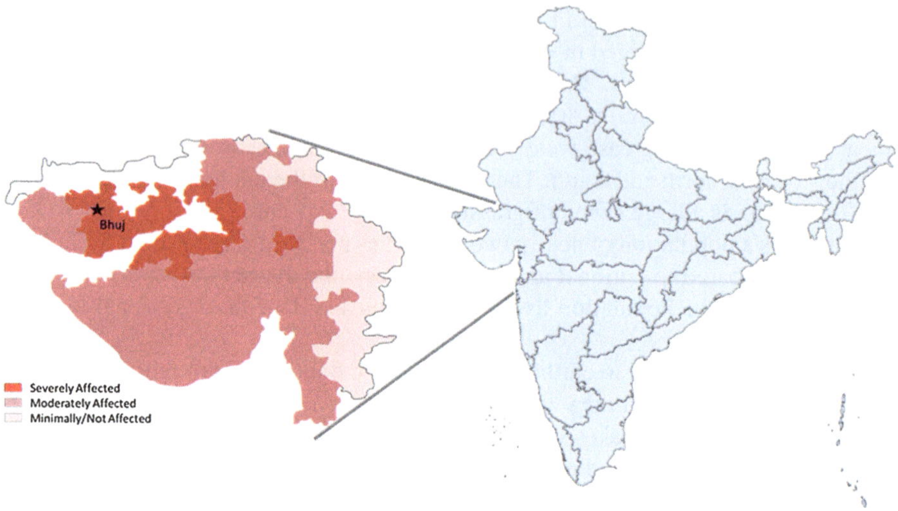

Fig. 2.2 Areas affected by the 2001 Gujarat earthquake (Source: Adapted from GSDMA (2001))

followed from 38 countries as well as United Nations agencies and many international NGOs such as the Red Cross. The national and state governments formed a group of special officers to handle the rescue, relief, and rehabilitation activities and mobilized funds for the same.

2.2 Capacity Crisis and Capacity Building After the 2001 Earthquake

2.2.1 Government and Donor Responses

Recognizing the need for coordination of government agencies, partners, and NGOs, the state government formed the Gujarat State Disaster Management Authority (GSDMA) in February 2001, just a month after the earthquake. The government also issued a preliminary report on earthquake damage assessment and invited donors to review the findings. Apart from physical impacts on buildings and infrastructure, the preliminary report mentioned the inadequate administrative structure to deal with relief and rescue work (Government of Gujarat 2001). As an initial response, the state government rushed in senior administrative officers to manage relief and rescue works. Later on, a stable higher-level administration was formed for longer-term relief, reconstruction, and repair by appointing additional collectors and additional district development officers (ADDOs) in the 16 worst-affected *talukas* (administrative divisions). These high-level officials were from an elite national Indian Administrative Service (IAS) cadre. At the apex level, a disaster management and mitigation authority, headed by the Chief Minister, and a disaster management task force to advise the government on relief and reconstruction policies and measures were set up. The task force was headed by a retired senior officer of the government with broad experience in relief operations.

The World Bank and the Asian Development Bank took the government's preliminary assessment as key input into a more detailed joint assessment carried out through field trips and consultations with government officers and NGOs. The joint assessment, which was finalized in March 2001, estimated sectoral asset losses to be US $2.1 billion for the same-standard replacement costs (including household assets) and US $1.8 billion in improved-standard replacement costs, excluding household assets (World Bank & Asian Development Bank 2001). The report proposed a recovery and reconstruction strategy based on (a) empowering individuals and communities by ensuring that the majority of reconstruction efforts be undertaken by the community; (b) a clear, transparent, and participatory approach to assess wishes of villagers and cost of alternatives; and (c) communication and transparency through effective dialogue among the government, public, and partners. GSDMA capacity building needs were identified as (i) immediate needs, which included coordinating all agencies and stakeholders involved in reconstruction, providing the financial management of Gujarat Emergency Earthquake Reconstruction Project (GEERP), funding and monitoring progress of the overall program, and developing a comprehensive and sustainable disaster risk management program, and (ii) long-term disaster risk reduction needs, which included disaster risk mapping (building on the Vulnerability Atlas of India for Gujarat, 1997) for disaster scenarios and microzonation; risk reduction by reviewing existing preparedness measures at state, district, and community levels to identify gaps; and risk transfer through insurance schemes and access to quick finance during disasters.

2.2.2 *Design of Capacity Development Project*

A massive donor-supported comprehensive rehabilitation and reconstruction program—Gujarat Emergency Earthquake Reconstruction Project (GEERP)—costing US $1,765 million was planned based on the government and donor assessments and launched by the government of Gujarat. This was funded jointly by the state government, the government of India, and bilateral and multilateral funding agencies such as the World Bank, the Asian Development Bank (ADB), the Netherlands government, the United States Agency for International Development (USAID), and the European Union (EU). The GEERP was designed as a comprehensive multi-sector program, aimed at rehabilitation of people through provision of housing, social amenities, infrastructure, and livelihood support based on a sustainable economy and environment and preparing them to face disasters through community participation and multi-hazard preparedness programs.

The government of Gujarat developed the GEERP as a comprehensive multi-sector program, aimed at rehabilitation of the people affected by the earthquake through provision of housing, social amenities, infrastructure, and livelihood support and based on principles of sustainable economy and ecology (GSDMA 2001). The project had three phases with the short- and medium-term phases focusing on recovery and reconstruction and the long-term phase focusing on capacity building for disaster reduction. The overall objectives of the long-term phase were to implement a comprehensive disaster management program and improve the disaster preparedness and emergency response capacity of the government to deal with different types of disasters.

The strategic focus of GEERP is shown in Fig. 2.3. The project budget was US $1.7 billion, out of which the state government secured a loan of US $687.5 million from the World Bank, US $350 million from the Asian Development Bank (ADB), and the rest from the national government and other state governments. The World Bank funding focused on housing, the social sector, infrastructure, community participation, and disaster management capacity building. The ADB funding focused on housing, urban/rural infrastructure, power, livelihood support, and disaster preparedness and mitigation. Other donors including the EU, USAID, Netherlands government, International Fund for Agricultural Development (IFAD), and World Food Program (WFP) also provided significant support by directly funding small activities.

The World Bank supported phase I of the project with a US $261.6 million loan, approved in March 2001, aimed at immediate reconstruction of housing, dams and irrigation, roads and bridges, and public buildings and support for health, education, and community participation. The second phase of the World Bank-supported lending provided US $442.8 million, approved in June 2002, for restoration of housing and public buildings, restoration of basic infrastructure such as the roads and irrigation sectors, and development of an institutional framework to allow better disaster mitigation and risk management for future natural disasters. One of the key objectives of the GEERP was to systematically enhance the capacity of government agencies and communities as well as to increase community involvement in managing reconstruction requirements and risk from future disasters (Mishra 2004; GSDMA 2006; World Bank 2009).

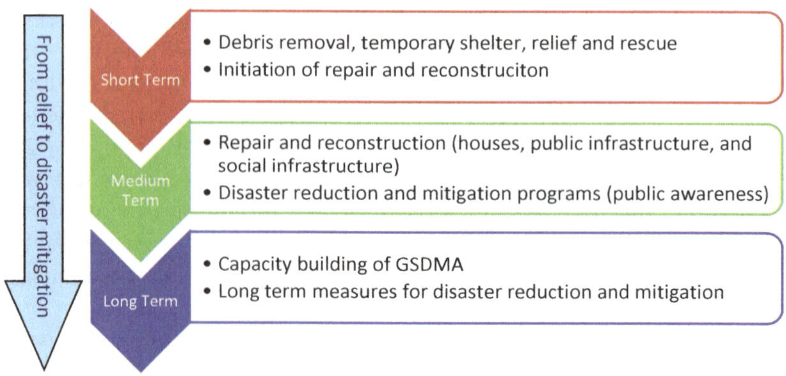

Fig. 2.3 Strategic focus of GEERP (Source: Adapted from Mishra (2004))

2.2.3 *Policy and Institutional Changes After the 2001 Earthquake*

Apart from the targeted program, GEERP, various policy, institutional, and funding changes occurred at the national and state level, providing needed legal backing and financing. The changes were especially welcomed by the donors and international aid agencies as it ensured sustainability of the results to be achieved under the GEERP.

2.2.3.1 National Level Changes

At a national level, National Disaster Act was passed in December 2005, 4 years after Gujarat earthquake and a year after Indian Ocean tsunami. The act paved a way for establishing National Disaster Management Authority (NDMA), responsible for preparing policies and plans for disaster management, and National Institute of Disaster Management (NIDM), responsible for research, training, and other capacity building activities. A National Executive Committee was formed under NDMA to implement policies and plans developed by NDMA.

A similar structure was suggested at state and district levels, with state and district disaster management authorities.

A National Disaster Response Fund (NDRF) and National Disaster Mitigation Fund (NDMF) were mandated by the 2005 Act, with similar funds at state and district levels.

2.2.3.2 Changes in Gujarat State

Within weeks after the 2001 earthquake, the state government in Gujarat set up a special body, the Gujarat State Disaster Management Authority (GSDMA), to deal with all aspects of relief and rehabilitation. Headed by an IAS officer, GSDMA worked with special officers at the district headquarters to coordinate relief activities in the early stages. This agency was conceived to be a permanent arrangement to handle natural disasters (see Sect. 5.1, subsection titled "Disaster management within the state administrative structure" for details about GSDMA). A Gujarat State Disaster Management Act was passed in 2003 to provide permanent status to GSDMA.

The Gujarat Institute of Disaster Management (GIDM) was established as a training and research wing of GSDMA on January 26, 2004, by the government of Gujarat with the aim of human resource development in the state. Its objectives include providing disaster management training, undertaking public education and community awareness, acting as a resource center and clearing house of information, and facilitating partnerships with private organizations and universities. Currently, GIDM offers a series of training courses to government officials and other stakeholders. These courses are offered by experts in the field. Four area development authorities were established in the four affected towns of Kutch to develop, coordinate, and implement urban development plans. These area development authorities are now responsible for providing building code permission and enforcement. Many of these functions were with local municipalities before 2001.

2.2.4 Is Kutch Any Safer than in 2001?

More than a decade later and with $1.7 billion spent in targeted capacity building program in Gujarat, primarily Kutch, is the region any safer? This question, although crucial, is very difficult to answer, especially in the absence of another earthquake. Should we wait till the next earthquake to find out?

At the "outside," looking at the external form, a phenomenal change has happened in the state. New policies and institutional structures have been adopted, which were up and running in a very little time after the earthquake. The GSDMA was established a month after the earthquake and has received many prestigious awards for its functioning. District[2] Disaster Management Agency, District Emergency Operation Center (DEOC), and Taluka Emergency Operation Center (TEOC) are all established and functional.

If you look closely at the affected settlements, however, the story is a little different. A survey conducted in 2011 found that 40 % of the surveyed buildings had high vulnerability to earthquake (Powell 2011). Some people were still living in temporary shelters in 2012 or have developed extensions to their houses—which were not earthquake resistant. Additionally the quality of construction has not improved even with masons training program. Another citizen survey conducted in 2012, as a part of this research in Bhuj, found that a majority of people surveyed believe that the town's capacity to deal with emergencies has improved but the next earthquake can have same impact as 2001, if not more (see Chap. 6). One reason for this belief was structural vulnerability of buildings in the city, which, along with the town's increasing population, puts more people at risk. Population in Bhuj, the capital city of Kutch district, has increased 49 % after the earthquake (World Bank 2009; Census of India 2011), while the city area has doubled, in part due to the economic incentives provided after the earthquake—exposing more people and assets to potential future earthquakes.

What is missing to make the massive investment more effective at the ground level? Is this a capability trap situation or just a typical capacity development process? How to measure results of capacity building in disaster risk management? What are the indicators of capability trap situation? How to break it? The next part of the book focuses on these questions, exploring what capacity building in disaster risk management means. A conceptual model will be developed to understand the Gujarat case.

---

[2] Somewhat equivalent to a county in the USA.

A detailed discussion on Gujarat's post-earthquake recovery and reconstruction program provides a great opportunity to understand the capacity building process in many ways. First, before the earthquake, the local administration followed an ad hoc approach to disaster management and paid little attention to disaster risk reduction. Soon after the 2001 earthquake, the GEERP was launched to systematically enhance government's capacity to manage reconstruction requirements and risk from future disasters. Second, new state, local, and national agencies were created with specific functions of dealing with future disasters. Third, after the earthquake many community-based organizations became very active in raising community concerns regarding relocation programs and the government's overall response to reconstruction and recovery. Finally, 6 years have passed since the donor-supported programs were completed,[3] giving adequate time to judge sustainability and effectiveness of the results achieved so far.

In assessing the success of the GEERP as a capacity development program, it is useful to envisage a potential failure situation. A capability trap may occur when even with organizations existing to deal with disasters, no real or relatively less capacity for preventing or preparing for disasters exists at the local level. Since one of the long-term goals of GEERP was capacity building for disaster reduction and mitigation, a capability trap situation would indicate an inability to achieve this goal.

References

Census of India. (2011). *Bhuj city census 2011 data.* http://www.census2011.co.in/census/city/317-bhuj.html. Accessed Dec 2012.

Government of Gujarat. (2001). *A preliminary report on the earthquake damage in Gujarat.* Gujarat: Government of Gujarat.

GSDMA. (2001). *Gujarat earthquake reconstruction and rehabilitation policy.* Gandhinagar: General Administration Department, GSDMA.

GSDMA. (2006). *GEERP quarterly progress report January–March 2006.* Gandhinagar: Gujarat State Disaster Management Authority.

Guha-Sapir, D., Below, R., & Hoyois, P. (2014). *EM-DAT: International disaster database.* Brussels: CRED. www.emdat.be. Accessed Mar 2014.

Mishra, P. K. (2004). *The Kutch earthquake 2001: Recollections, lessons and insights.* New Delhi: National Institute of Disaster Management.

NIDM. (2009). *Financing disaster management in India: A study for the thirteenth finance commission.* New Delhi: National Institute of Disaster Management.

Powell, P. J. (2011). Post-disaster reconstruction: A current analysis of Gujarat's response after the 2001 earthquake. *Environmental Hazards, 10*(3–4), 279–292.

UNISDR. (2009). *Global assessment report on disaster risk reduction.* Geneva: United Nations.

World Bank, & Asian Development Bank. (2001). *Gujarat earthquake recovery program: Assessment report.* http://www.preventionweb.net/files/2608_fullreport.pdf. Accessed Apr 2012.

World Bank. (2009). *Implementation completion and results report on Gujarat Emergency Earthquake Reconstruction Project.* Washington, DC: The World Bank.

[3] World Bank-supported GEERP was initiated in November 2001 and ended in October 2008. The original closing date was 2005; which was later revised to 2008.

Part II
Effective Capacity for Managing Disasters

Chapter 3
Understanding Capacity Development

Abstract The practice of capacity development is considered a "silver bullet" or a "cure all" by development scholars and practitioners. The concept, however, falls short of achieving any meaningful changes on the ground. In practice, the concept is applied without understanding how actual capacity building takes place, how institutions evolve, and how to define development in the first place. It is thus no surprise that the concept is now becoming associated with quite the opposite of what it intends to do. The scholars are now referring to a "capability trap" and further hollowing of capacity as the concept is applied in a rushed, top-down, and supply-driven manner. Though capacity building has different meanings to different groups, scholars and practitioners have developed variables and indicators to describe different aspects of capacity building. The literature suggests that there is no systematic framework to help scholars and practitioners understand and measure sustainable capacity development. This chapter provides a brief overview of the current literature and debates on capacity building within the larger landscape of international development practice. The chapter is divided into three main parts: (i) definitions and historical evolution of the capacity development concept, (ii) main research areas, and (iii) gaps in the current research.

Keywords Capacity development • Capability trap • Best practices • Top-down • Bottom-up

The practice of capacity development is considered a "silver bullet" or a "cure all" by development scholars and practitioners. The concept, however, falls short of achieving any meaningful changes on the ground. In practice, the concept is applied without understanding how actual capacity building takes place, how institutions evolve, and how to define development in the first place. It is thus no surprise that the concept is now becoming associated with quite the opposite of what it intends to do. The scholars are now referring to a "capability trap" and further hollowing of capacity as the concept is applied in a rushed, top-down, and supply-driven manner. Though capacity building has different meanings to different groups, scholars and practitioners have developed variables and indicators to describe different aspects of capacity building. The literature suggests that there is no systematic framework to help scholars and practitioners understand and measure sustainable capacity development.

A. Tiwari, *The Capacity Crisis in Disaster Risk Management*, Environmental Hazards, DOI 10.1007/978-3-319-09405-2_3

This chapter provides a brief overview of the current literature and debates on capacity building within the larger landscape of international development practice. The chapter is divided into three main parts: (i) definitions and historical evolution of the capacity development concept, (ii) main research areas, and (iii) gaps in the current research.

3.1 Definitions and Evolution of the Capacity Development Concept

Simply put, capacity is the ability to achieve a desired purpose (Brinkerhoff and Morgan 2010). Capacity development is thus enhancing the ability to achieve a desired collective purpose. Capacity development has a long history in international development management and with time has become a buzzword to include different types of development assistance from donor countries to developing and poor countries. Capacity development remains an elusive or imprecise concept as it can mean different things to different people (Kaplan 2000; Straussman 2007). Almost each international development agency has its own definition of what capacity development is and what it means in practice (see Table 3.1). While many donors such as the United Nations Development Programme (UNDP), the Deutsche Gesellschaft für Technische Zusammenarbeit (GTZ), the Canadian International Development Agency (CIDA), and the Organisation for Economic Co-operation and Development (OECD) agree that capacity development refers to capacity of individuals, organizations, and societies, there is relatively little agreement on the end goal of capacity development (UNDP 2008). The majority of donors believe that the end goal of capacity development is to achieve their own development objectives (UNDP 2008); others also believe that capacity development should lead to "transparent and accountable governance," "help in formulating and implementing policies," "effective use of resources," "improved performance" (Whyte 2004), and "enhanced problem-solving abilities" (OECD 2006). The World Bank's recent capacity development definition is very unique as it refers to capacity development as a "locally driven process of transformational learning by leaders, coalitions and other agents that leads to actions that support changes in institutional capacity areas – ownership, policy, and organizational – to advance development goals" (World Bank 2011).

The concept of capacity building evolved in the post-World War II era as a means to build and improve the physical and technical capacity of developing societies. Initially, in the 1950s, donor funding was used for nation building or physical capital. However, this kind of donor support did not show progress and so the focus shifted to capacity development, with an emphasis on human resource development and training in the 1970s and 1980s; policy reform, organizational development, and participatory approaches in the 1990s and 2000s; and poverty alleviation (e.g., millennium development goals), participation in capacity building, and

Table 3.1 Donor definitions of capacity development

Agency	Definition
UNDP	Capacity building is the process through which individuals, organizations and societies obtain, strengthen and maintain the capabilities to set and achieve their own development objectives over time
CIDA	Activities, approaches, strategies, and methodologies which help organizations, groups and individuals to improve their performance, generate development benefits, and achieve their objectives over time
European Union	To develop and strengthen structures, institutions and procedures that help to ensure: transparent and accountable governance in all public institutions; improve capacity to analyze, plan, formulate and implement policies in economic, social, environmental, research, science and technology fields; and in critical areas such as international negotiation
GTZ	Process of strengthening the abilities of individuals, organizations and societies to make effective use of resources, in order to achieve their own goals on a sustainable basis
JICA	The ongoing process of enhancing the problem-solving abilities of developing countries by taking into account all the factors at the individual, organizational, and societal levels
OECD	The process by which individuals, groups and organizations, institutions and countries develop, enhance and organize their systems, resources and knowledge; all reflected in their abilities, individually and collectively, to perform functions, solve problems and achieve objectives
World Bank	A locally driven process of learning by leaders, coalitions and other agents of change that brings about changes in sociopolitical, policy-related, and organizational factors to enhance local ownership for and the effectiveness and efficiency of efforts to achieve a development goal

Source: Whyte (2004), OECD (2006), JICA (2007), UNDP (2009), World Bank Institute (2009)

ongoing learning since 2000. The focus on capacity development programs led to an emphasis on new technical skills for government staff. As an extension to this, in recent times, international agencies have started taking a holistic view, and capacity development programs now try to develop systems that deal with administration, organizations, skills, technical and financial resources, and good governance. More recently, these agencies use the term capacity development, capacity enhancement, or capacity strengthening since capacity building can imply that there is no capacity to begin with.

Four streams of research can be identified in the capacity development literature. First, a number of donors and researchers have examined the performance of capacity development programs so far. The research in this stream paints a mixed picture. The second stream of research focuses on how capacity development occurs today: what are the current methods and tools to achieve capacity development? Many researchers have contributed to this stream of research, which is full of contentious debates and dilemmas on the value of current approaches to capacity development such as best practices and top-down or bottom-up, rapid or slow, and demand-driven or supply-driven donor programs. The third stream of research focuses on what is missing to make current capacity development programs successful. The question researchers ask in this strand is about what is missing or can be added to make effective changes. Debates in this stream focus on the demand side of capacity development and characteristics particular to specific country contexts. At the heart of this research is the idea that every country is unique and needs specific institutions and processes peculiar to that country's setting. The fourth stream of research focuses on normative models of capacity development—on how capacity development program should work to develop sustainable change. A small set of researchers have contributed to this field of research and provided piecemeal or partial understanding of the capacity development process.

3.2 Capacity Development Approaches

Capacity development programs are generally undertaken with the objective of achieving sustainable socioeconomic progress in developing or transition countries. However, in practice, capacity development activities have not always led to capacity improvements on the ground. Almost half a century has passed since the concept of capacity development was proposed. Still the whole international development field, in general, and the capacity development concept, in particular, have not radically changed the lives of people in developing countries. Many studies have shown improvements in individual skills after participation in capacity development programs, but sustainable organizational change has been rare (World Bank 2005; Grindle 2007; UNESCO 2009). Studies on development aid in Africa, for example, have shown that development or capacity deficits still exist (Easterly 2008a; Moss et al. 2008) even after relatively large donor funding for capacity development projects. It is in fact becoming difficult to break the vicious cycle of poverty leading to less capacity, which can again lead to greater poverty in Africa. In the words of Easterly, the researchers in Africa "see little sign of effect of these Herculean efforts at making civil servants perform better, even seeing some signs of decline" (Easterly 2008a, p. 43). Many fragile and poor nations seem to be stuck in this stage, even with a greater number of capacity development programs.

Donor evaluations also see little effect of large capacity building programs (World Bank 2005; Birdsall 2007). The World Bank's (2005) Operations Evaluation Department (OED) mentions:

> …the reasons for weak public sector performance are deeply rooted. Whatever pragmatic steps can be taken under the rubric of capacity building can be only a small part of the solution. Yet there is little empirical evidence to clarify what part of the problem international capacity building support can best help to solve; in what order capacity needs should be addressed; what can be expected of different kinds of interventions and why; and how knowledge of such processes as organizational change, learning, and incentives should shape capacity building efforts. (pp. 6–7)

Going a step further, some researchers have cautioned that external aid, if not used properly, not only risks being ineffective but can in fact contribute to the vicious cycles of a "capability trap," in which capacity remains low, and a "growth trap," where the business community of a country never gets a chance to thrive and government relies on external help. Decades of external aid in many aid-dependent countries in Africa show that if external aid is to be helpful for institution building in Africa's weak and fragile states, donors need not emphasize providing more aid and should minimize the risks more aid poses for this group (Birdsall 2007).

3.2.1 Current Capacity Development Tools and Approaches Leading to Development Mirage

One way to understand why international donor programs have not been successful is to understand the way capacity development through international aid actually works. While developing or underdeveloped countries want to "develop" and reach higher economic growth and Human Development Index levels, their means of achieving these ambitions can actually lead them in the opposite direction. Generally, leaders and bureaucrats in developing countries want to reach the status of developed countries very quickly but without the years of social struggle or technological innovation that have led to sustained social and economic transformations in high-income countries. This is in part due to the dreams shown by international development agencies and partners that such rapid transformations are possible through simple transfer of technology and best practices. These political leaders and bureaucrats, who are under immense pressure to show rapid progress to their people, are happy to accept this idea. Donors, on the other hand, in the absence of any proven capacity development practice, push certain technology and research in a supply-driven and top-down manner. While this "higher-level match" works for elites in the country including politicians, bureaucrats, and the wealthy class, it does not effectively trickle down to lower-level and common citizens. The result is an ever-increasing "development gap" between "wishful ambition" and reality as well as between the social and economic growth rates in high-income and low-income countries (Pritchett and Weijer 2010). Figure 3.1 below shows this process, labeled a "development mirage," where capacity development projects start with a solid promise of development and with high expectations but fail to provide desired results on the ground. In the mirage, the international development assistance, relying on best practices and supply-driven and short time-frame projects focusing on fiscal disbursement and short fixes, leads in the opposite direction.

The central approach to capacity development has been to graft best practices from developed countries to developing countries with the idea that what has worked in other places will work in any country environment. The artificial grafting is done in a top-down and supply-driven manner, on "form follows function" premise, i.e., if systems showing characteristics of developed countries are artificially grafted onto developing countries, functions will follow (Pritchett and Weijer 2010). However, the new organizations or systems have limited capacity since people and the society have not had time and training to transform and evolve into better and more organized systems. In addition to having little capacity to perform, the new

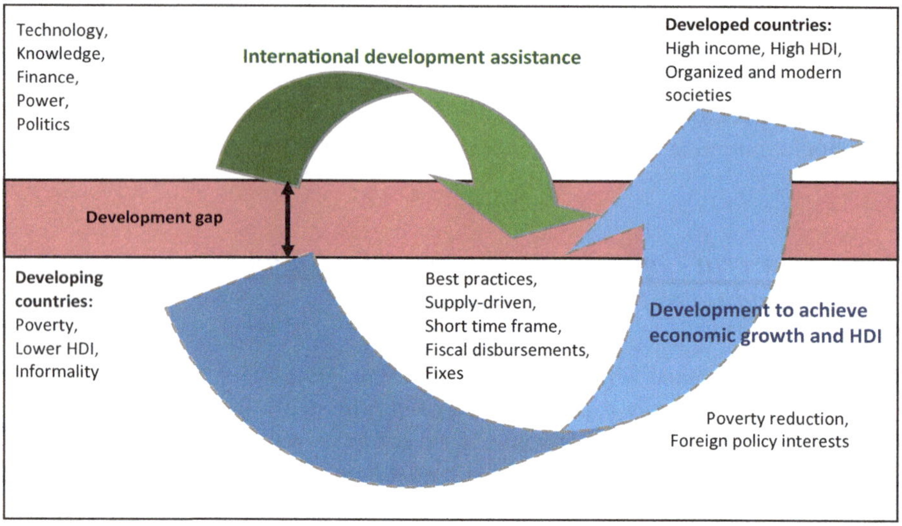

Fig. 3.1 Development mirage (Source: Adapted from Pritchett and Weijer (2010))

system or organization is expected to perform "too much of too little too soon too often" (since the organization or system "looks" ready to perform based on its form—"premature load bearing" in Pritchett and Weijer's terms). This actually crushes the system, leaving it in a "paralyzed" state where it mimics the form on the surface but remains weak underneath. With more of such programs, the entire developing country develops no sustainable capacity and becomes more and more aid dependent, which in fact worsens its situation. Pritchett and Weijer call this "big stuck" or "capability trap." Some of the current tools that are used for capacity development are discussed below along with the reasons why these tools create a "development mirage" instead of achieving positive and sustainable change.

3.2.1.1 Transfer of Best Practices

Development practitioners—the staff and consultants of international development organizations such as the World Bank, the Asian Development Bank, and the UN—design capacity development programs and projects that are implemented in recipient countries. Capacity development design and practice relies a great deal on what these practitioners think are best practices, which may not necessarily be derived from a proven theory or evidence. These "one-best-way models" and transfer of best policies and practices from developed to developing world are the predominant mechanisms currently adopted by development specialists to build capacity in developing countries. Straussman (2007) argues that capacity development forms a part of transnational dispersion, in which development specialists based in developed countries or matured economies provide advice to developing nations or transition economies. The advice given conveys ideas, norms, institutions, and practices from developed to developing nations.

However, the international development community agrees that the process of adapting best practices to developing countries is usually not successful. The most important factor is the questionable assumption that the best practices are really the best. A practice that is successful in one place need not be so in another place. There are various critical differences between the two places, especially when they are in a developed country in the West and a developing country. There are cultural, social, institutional, and political differences between countries. Even if this was not the case, the idea of replicating "best practices" has limitations. The "best practice" occurred in the past. In trying to emulate a past action, they are not looking to the future. In effect, they are trying to catch up. Where is innovation in this? On the extreme, one can even question the fundamental idea of best practice. How can one action be the best practice for anyone else unless both parties have the same goals and objectives? We can draw parallels here to capacity development programs. What is a right, appropriate, and forward-looking program?

Andrews (2008) argues that a good government means different things in different developed countries. Thus developing a "one-best-way model" (based on Sweden or Norway) can lead to "isomorphism, institutional dualism, and flailing states and imposing an inappropriate model of government that kicks away the ladder today's effective governments climbed to reach their current states" (p. 2). Isomorphism is the tendency of organizations to imitate characteristics that are generally considered effective. This could also be considered as the often-used terms "best practices," "good practice," or "international standard." Institutional dualism refers to a situation where "well intended legal, regulatory and procedural changes…" lead to governance that has "little bearing on how public decisions are actually made and implemented. Meanwhile, already existing understandings and practices continue shaping the way people are ruled" (Brinkerhoff and Goldsmith 2005, pp. 199–200). In practice, this all means that overlapping or contradictory systems are developed on top of existing formal or informal systems in developing countries, which ultimately leads to failure.

3.2.1.2 Top-Down Approaches

Capacity development programs are designed in a rapid and *top-down* manner without respecting the informal or evolutionary bottom-up process going on in a particular country. Notably, Easterly (2008b) offers some perspectives on how to understand this process within developing countries. He argues that most donor-funded programs assume a top-down view that institutions can be changed or developed primarily from the top by political leaders or the administrative system of a country. There is a formal hierarchical political and administrative system. The central/federal government controls most of the power at the top while local municipal bodies will be at the bottom of the ladder. People's perspectives may not have any strength. Financial and other powers diminish as we climb down the ladder. This may work well in an authoritarian system. On the other hand, many countries might already have informal systems of managing development that are *bottom-up*. In these systems, local communities will have a lot of say in what happens in the neighborhood, and bringing together such neighborhood actions will constitute larger plans and programs. The top-down and rapid programs tend to overlook bottom-up or evolutionary processes going on in these countries. This may lead to a fatal situation where informal institutions die and top-down institution building processes fail, as they are too fast or artificial in that country. Thus the recipient country ends up being worse off than before. Many informal or bottom-up institutions are already working in these countries. An attempt to overlook them can have negative consequences for the overall success of capacity development programs. Fortunately, many countries are now moving toward a more decentralized system, often with the support and encouragement of Western donor countries.

Easterly (2008b) and Pritchett and Weijer (2010) argue that developed countries have gone through a gradual evolutionary institution building process over many years before reaching today's matured state. This process has been "organic," unaltered by any outside or artificial impetus. The conditions in many developing countries are comparable to those in the USA 100 years before. In those days, many US cities did not have a safe drinking water supply or a participatory development process. Over many decades, those were developed after many ups and downs. That organic process and involvement of communities and institutions ensured that systems are developed and sustained. This creates a sense of ownership and pride in the people as well. Thus the resulting institutions are unique and appropriate for a country's situation, responding to historical, cultural, and other country-specific socio-economic and political characteristics. Today, we are trying to copy these institutions and transplant them to developing countries without any of the internal processes and adjustments the Western communities went through. It may be argued that one need not and cannot wait for so many years to bring development. However, the current development process expects all the countries to easily transform to the current levels of Western development in an instant, which may not be realistic. Overlooking this fact, development professionals still want to import ideas and practices from developed to developing nations, with the aim of quick fixes and rapid growth ["wishful thinking" in the words of Pritchett and Weijer (2010)].

3.2.1.3 Supply-Driven Approaches

The majority of capacity development support programs are primarily supply driven. That is, rather than being driven by the need or demand of the developing country, the donor-assisted capacity development programs move from one international development fad to the other, starting from infrastructure and going to agriculture, social services, governance, and poverty reduction (World Bank 2005). The international development agencies and most Western countries have their own agenda and approaches to development. They have money to disburse for particular items that they think are most pressing. On the other end, the developing country involved might have other pressing needs for which it is looking for help. However, since the funds are available for only a certain issue, the country is forced to accept money and work on that. This is where the disconnect between supply and demand exists. Recent donor focus on climate change mitigation and adaptation programs is a good example of this point. While many developing countries are still struggling with the delivery of basic services, the international focus on climate change means that a lot of donor support is now diverted to developing capacity to manage climate change in developing countries. Even with technical and financial grants, however, development practitioners find it difficult to undertake climate change-related programs in developing countries as there is little demand for such programs until basic services or institutions are institutionalized. This is pretty much like providing salt and pepper to a hungry person instead of a nutritious meal. Not surprisingly, climate change-related capacity development programs have not taken off successfully.

Why do development professionals follow international development fads, rather than trying to understand and fix the root cause of underdevelopment? Of course, the availability of donor funding in the new fields guides these changes. Another important reason is that development agencies do not want to be involved in "sensitive issues" such as politics, religion, and culture or tradition, which may be the root cause of the problem or underdevelopment. This could be, partly, because of the lingering effects of a Cold War mentality. In the 1960s, 1970s, and 1980s, Western countries tried to extend financial support to stop Soviet influence from spreading while the Soviet Union did the same to stop Western influence. Thus, while some countries were favored, others were neglected. This is particularly true for those developing countries that had geopolitical importance. In the post-Cold War scenario, the political landscape in many countries is different. While some countries have a matured or maturing democracy, many others still have military or authoritarian regimes. Religion also plays an important role. However, dealing with the intricacies of religion could be very controversial and development agencies try to avoid this. Culture and traditional values in some countries also play similar roles in forcing agencies to avoid dealing with them.

The nonsensitive sectors, where development professionals work, thus remain areas for technical skill-building programs and technology transfer that do not aim at the root cause of the problem. Administrators (especially those on job rotation) and politicians have shorter time frames to show results. So they also sometimes "muddle through." Politicians want to show positive changes on the ground within 4–5 years, which is usually the term of office. If they do not show that they have made changes, they may not be reelected. This means that they often show greater interest in programs or projects with short time frames, avoiding long-term programs that have the potential to bring fundamental changes in the structure and processes. Similarly, administrators have a rotation policy of 3–4 years. They will move from their current location or position within that time frame. As a result, either they do not take up long-term actions or do not follow up on the actions taken up by the previous officer. Political interference with day-to-day administration is another major hindrance for these officers. Both parties are thus less interested in long-term behavioral changes than in short-term fixes and results gained through technology transfers and skill-building programs.

3.2.1.4 Scope of Capacity Development Programs

The literature also points to the focus of capacity development efforts on the scope of the program rather than on implementation. At the same time, largely due to the pressure to perform and show results in a shorter duration, programs that have a short time frame are chosen over those with a long time frame. However, this does not give enough time to adjust the "process of change," and so gradual transformation is difficult. This leads to reliance on form over function, all of which can lead to "isomorphic mimicry, wishful thinking and premature load bearing" [explained earlier as conditions of "big stuck" (Pritchett and Weijer 2010)].

Debates over the right balance of large-scale top-down development and capital projects or small-scale bottom-up approaches have also not yielded any useful results. While large-scale projects are important for development, they are often disconnected from the community and may be unsustainable in the long run. Small-scale development seems ideal to enhance innovation and involve the community through bottom-up approaches. However, in practice, it can lead to "elite capture" and the risk of being small and non-scalable. Rao and Ibanez (2005), based on a review of participatory and community-driven social investment funds in Jamaica, showed that the actual investments were primarily aligned with the ambitions of local elites. Other approaches to capacity development, such as purely bottom-up approaches, which focus directly on helping communities and ignore the state, have also been unsuccessful in bringing sustainable change (Pritchett and Weijer 2010).

3.2.2 *Missing Success Ingredients*

Dilemmas related to capacity development approaches discussed above raise important questions for development efforts: how did developed countries reach where they are; what are the ways in which bottom-up processes are happening in developing countries; and how can we understand the various contexts in which the reform will or should take place? The literature in these areas is rather thin. Brinkerhoff (2010) argues that in the midst of many dilemmas and debates over how to get institutions and governance corrected or built in developing countries, it is easy to forget that capacity development is an internal or endogenous and dynamic process within a country. Irrespective of what donors do, the country's internal dynamics affects its capacity to start, implement, and sustain a reform. These internal dynamics include social, political, and economic factors. The societies in most developing countries are plural and often have hierarchical structures. India has such a diverse society that people in each state speak a different language, have different eating and dressing habits, and have different social value systems. There is a very rigid hierarchical system of power and authority that is followed even today. The educational and health conditions of the society will have a huge influence on what can be achieved. A highly educated society might have a better capacity to understand complex ideas, conceptualize new ideas, and sustain them over long time. These social dynamics are reflected directly and indirectly in the administrative systems.

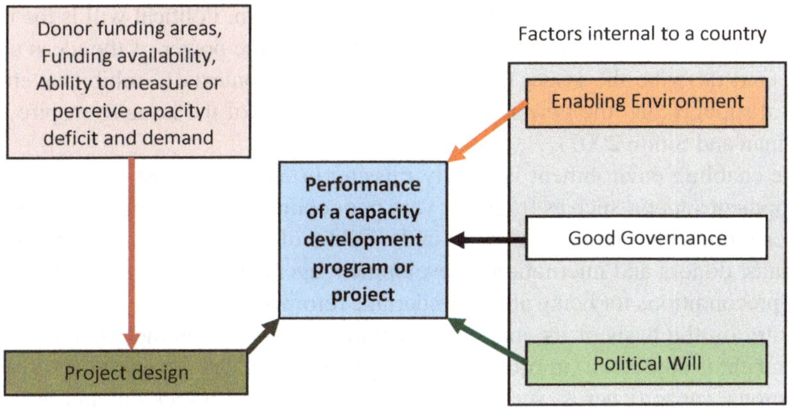

Fig. 3.2 Internal factors influencing the success of capacity development programs

Similarly, the political situation in a country has a lot of influence over the capacity of administrative systems. There could be vast differences in the systems of a country with an authoritarian regime from those in countries with vibrant multiparty democratic political systems. It is very important to recognize these dynamics to understand how a system works and how it can be improved.

However, instead of understanding the capacity development process, we see that the international development literature is full of many new ideas and concepts on what is missing to make capacity development successful. Three key concepts that relate to factors internal to a country stand out in this research: enabling environment, good governance, and political will (see Fig. 3.2). Similar to the term capacity development, enabling environment is also a popular term in international development agency circles, with a range of meanings from socioeconomic development itself to simply having a legal framework in place (Brinkerhoff 2010).

A general definition of enabling environment is:

> …a set of conditions - often inter-related - that impact on the capacity of citizens and civil society organizations to engage in development processes in a sustained and effective manner, whether at the policy, program or project level. They include legal, regulatory and policy frameworks, and political, socio-cultural and economic factors. There are institutional factors within civil society that should be considered in thinking about this environment. (World Bank 2011)

Good governance is another important concept in this debate. However, this concept is also very complex and differs greatly in its meaning. While governance deals with institutional process and rules of the game for authoritative decision-making (Grindle 2007), good governance is a more normative concept dealing broadly with stable macroeconomic policy, poverty reduction, decentralization, openness to trade, strong legislative bodies, and participation in decision-making. Grindle (2007) has rightly pointed out that it becomes hard to distinguish good governance from development itself. Political will is the next important concept, increasingly used alongside donor programs, in general, and capacity development programs, in particular. Especially in the context of millennium development goals, political will to initiate and sustain a reform is necessary for the success of a donor-supported program. Political will is the role of power—the power of actors connected with the issue, the power of the ideas used to define and describe the issue, the power of political contexts to inhibit or enhance political support, and the power of some characteristics of the issue to inspire action (Shiffman and Smith 2007).

The enabling environment is largely missing in countries most needing capacity development support such as fragile or very poor countries. Realizing that the absence of an enabling environment can result in the failure of capacity development projects/programs, donors and international development agencies have come up with a long list of preconditions for being able to undertake reforms. Increasingly, donors evaluate a country on the basis of its enabling environment. They examine whether national government is able to: (a) improve policy, legal, and regulatory frameworks; (b) build institutional capacity across sectors and at various levels; (c) seek out and respond to citizens' needs and preferences; (d) establish and maintain a range of oversight, accountability, and feedback mechanisms; and (e) mobilize and allocate public resources and investments (Brinkerhoff 2010). If these conditions exist, then a country will be eligible for grants or loans. One example is the USAID Millennium Challenge Account, which rewards countries that have put in place conditions for sustainability that will assure effective utilization of foreign assistance dollars. The World Bank and other institutions have gone even further and not only make the presence of these conditions critical for development lending but also monitor how countries perform (having these enabling conditions and performing better in post-program impact assessment) and make performance a precondition of future loans.

There is little research so far on the importance of developing an enabling environment through capacity development interventions and how that might be achieved. One can argue that developing an enabling environment is out of the scope of donor involvement and that it may be too broad a goal for a capacity development project. At the same time, there are not many methodologies or proven theories that can direct development professionals in designing a capacity development program that supports an enabling environment. Some of the important questions are as follows: (i) What is required to support enabling environment? (ii) How will development professionals know that they are doing enough to support an enabling environment? Current literature offers little on these questions, especially in light of unclear and fuzzy concepts related to capacity development and enabling environment. Grindle (2007) has already pointed out that preconditions may or may not lead to success. What is important is to have time and flexibility for the recipient countries to adjust to the reform.

3.2.3 Conceptual Models

Recently, researchers have been trying to develop a theoretical understanding of capacity development process. From the start of the twentieth century, we see attempts to understand the capacity development process, although the conceptual models so far are still normative—geared toward what should be done to increase capacity in poor nations—without a better understanding of what actually happens on the ground. Table 3.2 below provides an overview of current models of capacity development. Grindle (2007) provides a useful simple schematic for understanding the process, actors, and actions related to reform. Her model has three main elements: (i) *context* to assess what capacity currently exists, (ii) *contents* of proposed interventions, and

Table 3.2 Overview of current conceptual models on capacity development

Author	Characteristics	Comments
Grindle (2007)	Three main components:	Simple schematic to understand process, actors, and actions
	Context: two analytical frameworks to assess context or what existing capacity exists to build upon: (i) one to assess strengths and weaknesses of states and (ii) another that shows what sources of change might exist in particular environment	Not enough guidance on how community and civil society contribute to the process
	Contents: how easy or difficult proposed interventions can be and how likely are they to achieve their objectives in terms of conflict, time, organizational and logistical complexity, and behavioral changes	
	Process of reform: five "arenas" or stages of process in which characteristics of context and content come to play: (i) agenda setting or context, (ii) program/project design, (iii) bureaucratic and political adoption, (iv) implementation (based on characteristics of implementers, interests affected, intergovernmental structures, capacity of public sector, and new interests), and (v) sustainability (shaped by new stakeholders, implementer incentives, capacity to advocate and alliances)	
Brinkerhoff (2008)	Three levels—individual, organizations (group of individuals coming together for common purpose), and institutions (the rules, policies, laws, customs, and practices that govern how societies function). Three dimensions—time, degree of complexity, and the magnitude of change necessary. The capacity development interventions can be targeted at gaps and weaknesses in the following order: (i) resources (who has what), (ii) skills and knowledge (who knows what), (iii) organization (who can manage what), (iv) politics and power (who can get what), and (v) incentives (who wants to do what)	Largely top-down model focusing on capacity development targets
Grauwe (2009)	Four contextual levels in which capacity development takes place, starting from the background or system context within which public administration, organizational units, and individuals work. A number of issues that should be considered in each level for capacity development	Top-down model of and various levels in which capacity development takes place

(continued)

Table 3.2 (continued)

Author	Characteristics	Comments
Pritchett and Weijer (2010)	The middle way out, which relies on many leaders who are networked together to achieve clear outcomes on the ground (real capability), individuals who learn over time to achieve results together, and fiduciary accountability that provides flexibility for innovation while reducing the risk of corruption. Some other characteristics are (i) open systems that allow for earning, (ii) pressure for performance, (iii) organizations legitimated through demonstrated success, (iv) leaders focused on the creation of public value, and (v) frontline agents empowered to respond with concerned flexibility	Learning, leaders, and innovation required for sustainable change Many components are not clear, especially how they fit together and how they explain behavior

(iii) the *process of reform* in which characteristics of context and contents come into play. Grindle's model captures a number of important characteristics of reform initiation, design, adoption, implementation, and sustainability. The framework is as close to reality as can be represented in a schematic manner, and it provides a good start in understanding the whole process. However, a number of questions remained unanswered: What is civil society involvement? Is top-down or bottom-up, small or big, quick or time-consuming pace most useful for sustaining results?

Brinkerhoff (2008) has proposed a largely top-down model focusing on capacity development targets. His model comprises three levels—individual, organizations, and institutions—and three dimensions, time, degree of complexity, and the magnitude of change necessary in which capacity development takes place. He proposes that capacity development intervention should target identified gaps in resources (who has what), skills and knowledge (who knows what), organization (who can manage what), politics and power (who can get what), and incentives (who wants to do what). Brinkerhoff's model shows what capacity development can accomplish for an organization over time. However, his model is very simple, without much consideration for factors internal to a system such as the enabling environment, governance, political will, and civil society, which will play a key role in shaping the capacity development program and getting desired results. Although Brinkerhoff mentions that in-depth knowledge of country contexts would help in designing the capacity development effort, his model does not capture how this will be done nor does it shows the dynamism of the political and social forces that surround the capacity development effort.

Grauwe (2009) captures some of the recent thinking on various dimensions of capacity development and what should be considered in developing a capacity development project. His model consists of four levels of imbedded systems, starting from the background or system context within which public administration, organizational units, and individuals work. The system context includes social, economic, and political aspects. At the administration level, different levels of autonomy and distribution of roles can have varying effects on capacity to deliver services. At the organizational unit level, internal management including coordination among different units, transparency in actions, and proper supervision and support can have a big impact. Human resource management becomes an important factor. Finally, at the individual level, staff members' responsibilities and skills need to be matched up with his/her qualifications and training. Grauwe's model helps in our understanding of different levels in which capacity development takes place, context and institutions, and interests. However, it does not capture very well the process of change itself, as well as what is the role of the community in capacity development.

A more recent contribution to capacity development models comes from Pritchett and Weijer (2010) who emphasize the importance of developing many leaders over "one champion" and learning over training. Training is about building systemic or organizational capability. Learning, on the other hand, produces usable knowledge within the organization, as people learn how to achieve objectives. They also warn against taking financial accounting as a form of accountability. Many external agencies such as the World Bank implement a strict fiduciary accountability that wards off corruption in countries with the least capacity. However, such an externally imposed system does not help to create institutions in fragile states that are domestically legitimate and accountable. In other words, it does not help in developing an "enabling environment." Pritchett and Weijer provide a new look at the capacity building process and capture much of the dynamism that may actually govern the process. However, their model does not provide enough explanation on how to reach the goals. For example, they talk about the need of an adaptor plug that can connect two systems of accountability—external agent driven and those evolving in the country. However, they fail to describe what this connector could be. They do not show how to ensure a capacity development project's focus on organizational learning and how to develop many leaders. Many suggested components are not clear, especially how they fit together and how they explain behavior.

One of the key challenges in this field is not what should be done but how does the capacity building process work? How can capacity building be achieved? What can be done to have successful and sustainable capacity development in the developing country context? These are some of the key questions that many researchers have failed to answer. None of the models have successfully captured the "bottom-up" aspect of capacity building. How do civil society and community contribute to capacity development? A better understanding of what actually happens after undertaking a capacity development program over what was "wished" in the beginning would help in our understanding of and ability to implement pragmatic programs or policies to bring real changes on ground.

3.3 Capacity Development Process

Many scholars have noted the lack of a comprehensive theory explaining how actual capacity building processes take place in reality, how institutions evolve, how organic development processes take place, and how development should be defined (Easterly 2008b; Pritchett et al. 2010). Pritchett et al. (2010) call for a renewed research agenda focused on understanding capacity development process. In their words:

> Inadequate theory of development change reinforces a fundamental mismatch between expectations and the actual capacity of prevailing administrative systems to implement even the most routine administrative tasks. (p. 1)
>
> This analysis gives rise to a policy research agenda focused on better understanding the conditions under which political space is created for nurturing the endogenous learning and indigenous debate necessary to create context-specific institutions and incremental reform processes. (p. 45)

Less attention on theory development and more on new actions may have resulted from the need for quick and measurable results required by political representatives and donors to showcase and track progress. However, a theory on how capacity development happens is needed to increase our understanding of what effective capacity means and how it can be initiated and maintained to bring meaningful reforms in developing countries. As Andrews (2008) explains:

> The lack of theory underlying current indicators…is partly the result of a general rush to measure effectiveness without a theory of effectiveness. I would thus suggest that we not focus on developing new indicators immediately but rather on building an understanding of theory to underlie such, so that if we identify certain services as key and certain contextual factors as influential, we understand why, and also so that we have some ability to explain how effective governments adjust their challenge focus and choose the processes to meet challenges. (p. 38)

Theories or ideas are important to the field of development management (Grindle 2007). First, there is a strong mutual relationship between how ideas can generate new practices and help in solving current development challenges and how practice can show unique insights to refine "ideas" or theories of development. Second, ideas are particularly valuable for fragile or developing country contexts where institutions and organizations are still evolving or are fluid and are more likely to adopt to a new idea or theory (however, that is not a reason to treat these countries as laboratory mice in order to test many ideas or theories). Finally, understanding why certain factors are successful in certain contexts can guide future generations of development practitioners in prioritizing actions to achieve anticipated results, without being able to reinvent the wheel. Theories on capacity development should not only focus on whether capacity development can effectively be planned in advance and supported by outside intervention (Brinkerhoff 2010) but also on how capacity development happens—both within formal administrative circles and at the community level. Bottom-up approaches to capacity development process have not received much attention in the literature. An understanding of how community involvement contributes to capacity development process and what is the community's role in sustaining capacity thus remains vital to the understanding of the overall process.

3.4 Capacity Development: From a Buzzword to a Comprehensive Theory of Reform

The concept of capacity development is essential to the idea of governance and development, particularly in the developing countries. Governments and development agencies have been promoting and implementing many capacity development programs over the past five decades. This is particularly applicable to the field of international development with the logic that one cannot achieve or sustain change without capacity. The idea evolved after donor funding directed toward financial support, infrastructure development, or human resource development alone did not result in overall development. The discouraging results forced the development assistance community to combine a number of physical, fiscal, and logistic activities and integrate them into the programs and plans of governments at different levels, contrary to the previous way of piecemeal targeting of capacity development as stand-alone programs.

Capacity development has different meanings to different people (Kaplan 2000; Straussman 2007; Easterly 2008b; Andrews 2008) and is associated with a range of activities such as training, skill building, policy reforms, and institutional development (Rodrik and Subramanian 2003; Ritzen 2005; Grindle 2007). The concept has become a buzzword and is now a critical part of a majority of donor support programs—promising "development" by enhancing the capacity to bring, manage, and sustain "development." The concept has great appeal today and it is easy to understand why—how could one achieve or sustain change without capacity? In reality, however, the concept remains a development fad and, like many of its cousins such as good governance, political will, and enabling environment, has remained elusive and obscure. The concept falls short of development practitioners' expectations to attain sustainable and positive change—providing little guidance on how to decide what development means in a particular country or local context, how actions should be prioritized within limited means and competing priorities, and what actions should be taken on the ground.

Sometimes capacity building programs can reverse some of the progress already made (Abraham and Platteau 2004; Rao and Ibanez 2005; Pritchett and Weijer 2010) leading to a "capability trap." Capability trap is a situation where development efforts lead to nothing more than a proliferation of institutions, jargons, and expenditures, with no real improvement in the capability of the state to implement policies and programs (Birdsall 2007; Moyo 2009; Pritchett and Weijer 2010). The concept of capacity development has failed to perform up to the expectations of scholars and international development experts (World Bank 2005; Birdsall 2007; Grindle 2007; Pritchett and Weijer 2010). Researchers point to certain "persistent failure mechanisms" triggered by the way capacity building programs are designed and practiced. Much of these observations have come out of the fact that the capacity building programs have not led to phenomenal change on the ground, and the programs have not changed the development paradigm in any fundamental way. Many donor-led capacity development programs have shown some immediate success but no sustainable change in the long run. Thus, one is forced to think if it is even worth using the concept anymore.

One of the main reasons for this failure is the lack of a universally accepted theory or model that explains how the capacity building process takes place in reality (Brinkerhoff 2004, 2010; Grindle 2007; Grauwe 2009) and what factors contribute to or sustain capacity (World Bank 2005; Shiffman and Smith 2007; Andrews 2008; Grauwe 2009). Less attention to theory development and more on new actions result from the need for quick and measurable results required by political representatives and donors to showcase and track progress. Scholars now agree that none of the concepts developed so far have fully solved the situation. In fact, they have added to the "development mirage" in which new concepts start with a solid promise of development and amidst high expectations when they are applied in practice fail to provide desired results. Researchers are thus interested in developing theoretical models that can bring together different streams of capacity development and increase our understanding of how capacity development process happens on the ground.

References

Abraham, A., & Platteau, J.-P. (2004). Participatory development: Where culture creeps in. In V. Rao & M. Walton (Eds.), *Culture and public action*. Stanford: Stanford University Press.

Andrews, M. (2008). *Are one-best-way models of effective government suitable for developing countries?* Faculty research working papers series RWP08-014. Boston: John F. Kennedy School of Government, Harvard University.

Birdsall, N. (2007). Do no harm: Aid, weak institutions and the missing middle in Africa. *Development Policy Review, 25*(5), 575–598.

Brinkerhoff, D. W. (2004). *The enabling environment for implementing the millennium development goals: Government actions to support NGOs*. Paper presented at George Washington University conference "The role of NGOs in implementing the millennium development goals", Washington, DC, May 12–13, 2004.

Brinkerhoff, D. W. (2008). The state and international development management: Shifting tides, changing boundaries, and future directions. *Public Administration Review, 68*(6), 985–1002.

Brinkerhoff, D. W. (2010). Developing capacity in fragile states. *Public Administration and Development, 30*(1), 66–78.

Brinkerhoff, D. W., & Goldsmith, A. (2005). Institutional dualism and international development: A revisionist interpretation of good governance. *Administration and Society, 37*(2), 199–224.

Brinkerhoff, D. W., & Morgan, P. J. (2010). Capacity and capacity development: Coping with complexity. *Public Administration and Development, 30*(1), 2–10.

Easterly, W. (2008a). *Can the west save Africa?* NBER working paper no. 14363. Cambridge, MA: National Bureau of Economic Research.

Easterly, W. (2008b). Institutions: Top down or bottom up? *American Economic Review, 98*(2), 95–99.

Grauwe, A. D. (2009). *Without capacity there is no development*. Paris: UNESCO.

Grindle, M. S. (2007). Good enough governance revisited. *Development Policy Review, 25*(5), 553–574.

JICA. (2007). *Summary of report: Towards capacity development (CD) of developing countries based on their ownership – Concept of CD, its definition and its application in JICA projects*. http://www.jica.go.jp/cdstudy/library/pdf/20071101_24.pdf. Accessed on 20 Sept 2012.

Kaplan, A. (2000). Capacity building: Shifting the paradigms of practice. *Development in Practice, 10*(3 & 4), 517–525.

Moss, S., Dale, N., Nyirongo, J., et al. (2008). *Local voices, global choices: For successful disaster risk reduction*. Geneva: ProVention Consortium.

Moyo, D. (2009). *Dead aid: Why aid is not working and how there is a better way for Africa*. New York: Farrar, Straus and Giroux.

OECD. (2006). *Applying strategic environmental assessment: Good practice guidance for development co-operation*. Paris: Organisation for Economic Co-operation and Development.

Pritchett, L., & Weijer, F. (2010). *Fragile states: Stuck in a capability trap?* Input paper for The World Development Report 2011. Washington, DC: The World Bank.

Pritchett, L., Woolcock, M., & Andrews, M. (2010). *Capability traps: The mechanisms of persistent implementation failure*. Working paper 234. Washington, DC: Center for Global Development.

Rao, V., & Ibanez, A. M. (2005). The social impact of social funds in Jamaica: A 'participatory econometric' analysis of targeting, collective action, and participation in community-driven development. *Journal of Development Studies, 41*(5), 788–838.

Ritzen, J. (2005). *A chance for the World Bank*. London: Anthem Press.

Rodrik, D., & Subramanian, A. (2003). The primacy of institutions (and what this does and does not mean). *Finance & Development, 40*(2), 31–34.

Shiffman, J., & Smith, S. (2007). Generation of political priority for global health initiatives: A framework and case study of maternal mortality. *The Lancet, 370*(9595), 1370–1379.

Straussman, J. (2007). An essay on the meaning(s) of 'capacity building' – With an application to Serbia. *International Journal of Public Administration, 30*(10), 1103–1120.

UNDP. (2008). *Capacity development: Empowering people and institutions*. New York: UNDP.

UNDP. (2009). *Supporting capacity development: The UNDP approach*. New York: United Nations Development Program.

UNESCO. (2009). *Guidebook for planning education in emergencies and reconstruction*. Paris: UNESCO.

Whyte, A. (2004). *Landscape analysis of donor trends in international development* (Human and Institutional Capacity Building: A Rockefeller Foundation Series, Issue No. 2). New York: The Rockefeller Foundation.

World Bank. (2005). *Capacity building in Africa: An OED evaluation of World Bank support*. Washington, DC: The World Bank.

World Bank. (2011). *Enabling environment for civic engagement*. http://go.worldbank.org/4JJJ1MKSZ0. Accessed on 12 May 2012.

World Bank Institute. (2009). *The capacity development results framework*. Washington, DC: The World Bank.

Chapter 4
Capacity for Managing Disasters

Abstract Capacity for managing and preventing disasters depends upon how disasters are defined or understood. Until the nineteenth century, the capacity to manage disasters was limited to the ability to undertake relief and rescue operations. Beginning in the early 1900s, social scientists laid a foundation for understanding disasters as the product of natural and social forces. Based on this understanding, the capacity to manage "disaster risk" required the ability to develop good models of disaster risk, planning to mitigate known disaster risks, and transferring risk of economic losses from likely disasters along with the ability to be prepared for disasters. However, even with the new understanding and increased efforts associated with disaster response and mitigation, monetary losses from disasters are increased globally. Social scientists now believe that we live in a "risk society" where "industrial-technical-scientific projects" produce unintended risks that are incalculable, uninsurable, and uncontrollable. The capacity to manage disasters now requires an ability to learn, internalize and make change, manage information flow and exchange, ensure flexibility and adaptability in the structure and functions of organizations, and coordinate work with multiple agencies and units to achieve a common purpose. This chapter provides a strong theoretical basis of what capacities are needed for managing disasters and who needs these capacities. This chapter is divided into four sections: (i) the need for capacity development in disaster risk management, (ii) key concepts, (iii) historical evolution of the field and where it is heading, and (iv) what and whose capacity.

Keywords Disaster risk • Disaster risk management • Resilience • Risk society • Disaster research

Capacity for managing and preventing disasters depends upon how disasters are defined or understood. Until the nineteenth century, the capacity to manage disasters was limited to the ability to undertake relief and rescue operations. Beginning in the early 1900s, social scientists, primarily from the sociology discipline, laid a foundation for understanding disasters as the product of natural and social forces (Cardona 2003; Wisner et al. 2004; Drabek 2005; Mileti and Gailus 2005; World Bank 2010). Based on this understanding, the capacity to manage "disaster risk" required the ability to develop good models of disaster risk, planning to mitigate known disaster risks, and transferring risk of economic losses from likely disasters

A. Tiwari, *The Capacity Crisis in Disaster Risk Management*, Environmental Hazards, DOI 10.1007/978-3-319-09405-2_4

through insurance and other financial mechanisms, along with the ability to be prepared for disasters through quick relief, rescue, and rehabilitation operations.

However, even with the new understanding and increased efforts associated with disaster response and mitigation, monetary losses from disasters are increasing globally. Social scientists now believe that we live in a "risk society" where "industrial-technical-scientific projects" produce unintended risks that are incalculable, uninsurable, and uncontrollable (Beck 2006). The capacity to manage disasters now requires an ability to learn, internalize and make change, manage information flow and exchange, ensure flexibility and adaptability in the structure and functions of organizations, and coordinate work with multiple agencies and units to achieve a common purpose. However, to date, a comprehensive theory of risk and resilience, especially from a multidisciplinary point of view, is still missing (Cardona 2003) to guide actions and to answer questions such as: What real changes are needed in government structure and functions? Does the community have a role to play? How should the information be managed? How can a "learning culture" be encouraged?

This chapter will provide a strong theoretical basis of what capacities are needed for managing disasters and who need these capacities. This chapter will be subdivided into four sections: (i) the need for capacity development in disaster risk management, (ii) key concepts, (iii) historical evolution of the field and where it is heading, and (iv) what and whose capacity.

4.1 The Need for Capacity Development in Disaster Risk Management

Although almost all development sectors of an economy require capacity building efforts to sustain capacity and keep up with new challenges, the capacity to manage disasters is far more challenging. First, the recent focus on preventing disasters and their impacts rather than reacting to them, as has been done historically, requires new ways of thinking. Second, unlike other sectors of development, disaster management cuts across many sectors and requires additional and redundant capacity to coordinate actions, respond in time, and sustain capacity for a long time after the disasters are over. Third, increased funding for preventing and preparing for disasters is leading to the application of "regular capacity building" tools, which have been ineffective thus far. This includes grafting of best practices; rapid development of new agencies tasked with disaster prevention, especially soon after a major devastating disaster; project management issues related to unrealistic time period, primarily for post-disaster reconstruction; and quick fixes to achieve measurable results.

The focus on ex ante or preventive steps to manage disasters comes from the recognition that just "reacting" to disasters is not working, which is evident from the increasing damages from disasters globally (Vos et al. 2010; World Bank 2010). A recent publication from the World Bank and the United Nations provides further support for describing the increasing number of disasters based on scientific research using a database of disaster impacts over last forty years. Donors find the "reacting to disasters" approach as costly and unsustainable and are thus supporting this new trend to focus on disaster prevention and preparedness activities. Largely due to this change in thinking and donor efforts, an international agreement was formed in 2005, in the form of the Hyogo Framework for Action (HFA), which aims "to substantially reduce disaster losses, in lives and in the social, economic and environmental assets of communities and countries by effectively integrating, in a coherent manner, disaster risk considerations into sustainable development policies, planning, programming, and financing at all levels of government" (UNISDR 2009a; INEE 2012). The HFA was adopted by 168 countries in 2005 and is a political agreement to increase countries' disaster risk management capacities. Thus, new donor funding as well as government budget is now available to develop capacity in the areas of disaster prevention and preparedness (UNISDR 2011; INEE 2012).

Disaster management cuts across many sectors and thus needs a new way of working. For example, while experts focusing on water supply may be challenged by the alarming population growth in a city of a developing country, their work is still manageable if there is enough technical understanding of topography and sources of drinking water, as well as financial resources to put in new waterlines. Disaster preparedness activities would need to be coordinated across many agencies with mandates in public works, land-use management, environmental management, early warning, water resource management, geology, and emergency response.

Since the expectations to perform and show results are very high, the current capacity development methods rely on best practices, quick development of new agencies tasked with disaster prevention, and project management based on measurable targets. However, such approaches have not shown desired results. If the current approach does not work in developing countries, there is a danger that it will become another "development fad." Already the climate change agenda has emerged as a new fad over and above disaster mitigation dialogue. In their pursuit of "quick change," the donors can discard disaster risk management field as not showing enough results and thus can move to the next "silver bullet," in this case climate change adaptation. Thus, it is all the more prudent now to understand how capacity building in disaster risk management is practiced today, what are the underlying concepts contributing to the tools and methods of practice, what are the current debates and gaps in literature, and what are the "recommendations" or ways to improve it.

4.2 Key Terms and Concepts

Disaster risk management is a young and evolving field, with new concepts and definitions. This section contains key terms and concepts of the field as they are understood or practiced thus far.

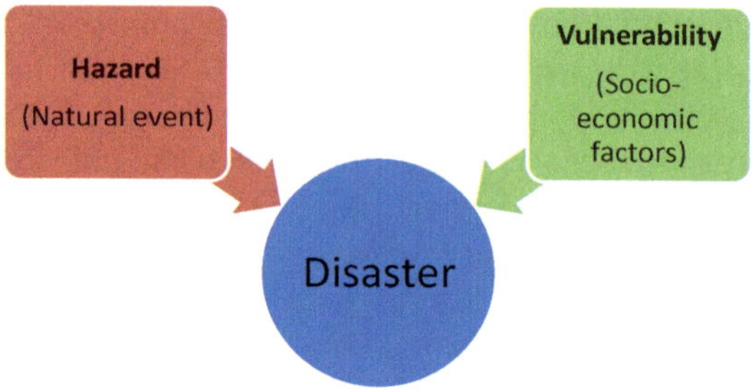

Fig. 4.1 Disaster as a combination of hazard and vulnerability

4.2.1 Disaster

A disaster is as an event that results in a great damage or loss of life. It is "a serious disruption of the functioning of a community or a society causing widespread human, material, economic or environmental losses which exceed the ability of the affected community or society to cope using its own resources" (UNISDR 2004, 2009b). A disaster is a specific event which is a result of a natural hazard meeting vulnerability (Wisner 2003; Drabek 2005; Mileti and Gailus 2005; Cardona 2003; World Bank 2010). Figure 4.1 above depicts this phenomenon. For example, New Orleans flooding in 2005 occurred when a natural hazard (in this case, Hurricane Katrina) was exposed to city levies, which could not withstand the impacts of the hurricane.

4.2.2 Disaster Risk

Related to the concept of disaster is the concept of disaster risk. While disasters are external shocks occurring as one-time events, disaster risk is internal to the development process and is always present. The concept of disaster risk is central to disaster prevention and is understood as a complex interplay between hazard, exposure, and vulnerability. A hazard, which is a latent danger of natural origin such as cyclones, translates into a disaster event where there is vulnerability. Damages and losses from a disaster depend upon the degree of exposure to social elements and their vulnerability. Disaster risk is multiplicative because for disaster risk to exist, all three components—hazard, vulnerability, and exposure—should be present (World Bank 2010).

4.2.3 Hazard

Hazard is a latent danger or an external risk factor (Cardona 2003; UNISDR 2004, 2009b) that represents potential harm to a community or an environment (Drabek 2005). It is a natural process or phenomenon classified by geophysical, meteorological, hydrological, climatological, and biological causes. Hazards are studied by physical sciences experts, who usually describe them through a severity scale and frequency. The distinction between natural and technological hazards can often be difficult. While many hazards can have a natural origin, man-made changes can exacerbate the frequency or intensity of the hazard.

4.2.4 Vulnerability

Vulnerability is defined as an internal risk factor. It is an economic, social, political, and physical susceptibility or predisposition of a community to damage in case of a destabilizing phenomenon of natural or ethnographic origin (Cardona 2003). While hazards represent physical conditions external to a system, vulnerability is internal to the system. It is always relative to a physical phenomenon; for example, a community may be vulnerable to droughts if its livelihood relies on rainfall, it is hard to store food for "bad days" due to poverty, and no external or government help is available to provide food and water in time to save people's lives, crops, or livestock. The same community may not be vulnerable to earthquakes because people are living in bamboo huts, which will not collapse during such quakes, and its livelihood, based on agriculture, is not affected by quakes.

The concept of vulnerability is crucial to understand how disasters are perceived today, and significant contributions have come from social science in regard to this field. Some social scientists have defined vulnerability as opposite to capacity or capability to prevent and respond to disasters. Wisner et al. (2004) defined vulnerability as characteristics of persons or groups in terms of their capacity to anticipate, cope with, resist, and recover from the impact of a natural hazard. Others have compared vulnerability to the ability to protect one's community, home, and family and to reestablish one's livelihoods (Anderson and Woodrow 1989). This definition of vulnerability as it relates to capacity provides a foundation for a capacity development agenda in disaster management and, thus, deserves a larger discussion.

The renewed focus on vulnerability results from social scientists calling for a "paradigm shift" since vulnerability can be controlled unlike hazards (Wisner 2003; Enarson et al. 2003; McEntire 2004). These scientists call for the goal of reducing vulnerability, not disaster damages, as, to them, vulnerability is the root cause of disaster. Given the sociological foundations of these arguments, they further believe that the root cause of vulnerability is the societal structure that favors those who have money and power and puts the underprivileged and weaker sections of the society at a higher risk of disasters. "Social systems generate unequal exposure to risk by making some people more prone to disaster than others and these inequalities in risk and opportunity are largely a function of the power relations operative in every society" (Bankoff 2003, p. 6). These social scientists ask, "Why must the patterns of greed and financial corruption continue to perpetuate so-called disasters wherein those most vulnerable are disproportionately hurt?" (Drabek 2005, p. 10).

Emphasis on vulnerability puts poor people at the center of disaster mitigation efforts as they are affected disproportionately by disasters due to many factors (Wisner et al. 2004). First, they are more likely to be living in hazard-prone areas. For example, a study of Bogotá, Colombia, shows that poor people tend to cluster in more hazardous areas (Lall and Deichmann 2009). Second, their housing and surroundings are less likely to be resistant to disasters, whereas wealthier people may be able to invest in disaster-resistant housing and are likely to have better access to infrastructure and services, such as a vehicle for evacuation and access to emergency care. Third, consequences of disasters are much greater for poor people as they have fewer resources to rebuild or recover, whereas the homes and assets of rich people are likely to be insured (not always the case in many developing countries). Finally, living in hazard-prone areas is most likely to be voluntary for rich people especially if the hazard threat is known. Disaster insurance and disaster prevention structures which can prevent disasters, such as levees, also provide incentives to live in hazard-prone areas. These measures often create a "false sense of security," thereby reducing smaller events but yielding catastrophic damages when these systems fail. However, for the poor, the choice to live away from hazard-prone areas may result in loss of livelihoods. For example, a study of slums in Jakarta shows that poor people take a conscious decision to stay in hazard-prone areas and face floods rather than moving out and losing their jobs (Budds et al. 2005).

Relatively recently, the capacity of institutions—policies, programs, and government machinery to prevent disasters, alert people, and respond in time—has also been considered as an important determinant of vulnerability (World Bank 2010). In many places, governments and organizations in charge of managing disasters do not work effectively because of the absence of political will, and so they focus on response. They work as centralized hierarchies and do not adequately incorporate the local power base such as municipal governments, community organizations, and civil society organizations (Cardona 2003).

4.2.5 Exposure

While many experts believe exposure to be a component of vulnerability, some consider it as a separate concept, especially those concerned with mathematical modeling of disaster risk (UNISDR 2009b). Hazard and vulnerability can exist together without resulting in disaster if there is no exposure. The degree of exposure or contact determines the degree of damage during disaster. For example, heavy rainfall will not lead to damages and losses if people and their assets are not exposed to flooding. The degree of damages and losses will depend upon how many people and assets come in contact with floods and how vulnerable they are to the effects of flooding.

4.2.6 *Disaster Risk Management*

Traditionally, the approach to managing disasters has included rescue operations during disaster and post-disaster relief and recovery/rehabilitation activities. Thus, terms such as crisis management, emergency management, and disaster management have been used with the demarcation of three distinct phases of disaster management: *pre-disaster*, being prepared for disasters; *during disasters*, response and relief; and *post-disaster*, recovery and reconstruction. However, in recent years, the term disaster risk management has gained widespread support with the realization that minimizing the disaster risk itself can significantly reduce the human and property losses. In this approach, mitigation and preparedness activities are always needed to control or manage disaster risk. The idea is not to think in "phases" since planning and preparedness activities are continuous processes, not goals to be accomplished and put aside (Dynes et al. 1972). Although the majority of countries still follow a "response" approach, new research is being conducted to test the benefits of investing in mitigation compared to response. For example, statistical analyses of the ex post and ex ante policy responses such as grain and livestock distribution in the aftermath of the 1994–1995 drought in Zimbabwe show that ex ante actions are welfare enhancing and poverty reducing (Owens et al. 2003).

4.2.7 *Resilience*

The term resilience is often used in the same manner as the notion of "bouncing back" that reflects its Latin root "*resiliere*" which means "to jump back." Holling (1973, 1986) defined the term resilience for an ecosystem as the measure of the ability of an ecosystem to absorb changes and still persist. Wildavsky (1988, p. 98) defined resilience as "the quality of managing surprise - the ability to absorb shocks gracefully." Walker et al. (2004, p. 1) updated Holling's definition of resilience as "the capacity of a system to absorb disturbance and reorganize while undergoing change so as to still retain essentially the same function, structure, identity, and feedbacks." Operationally, resilience implies a conceptual shift away from trying to create an "optimal" design of infrastructure, that tries to anticipate all shocks, toward recognizing that surprises are inevitable and public policy must reorient toward accepting uncertainty and design that is "good enough." In the context of cities, Alberti et al. (2003) defined urban resilience as the degree to which cities are able to tolerate alteration before reorganizing around a new set of structures and processes by measuring how well cities (and implicitly urban infrastructure) can simultaneously balance complex, tightly coupled, interconnected, and interdependent technological, ecosystem, and human functions.

4.3 Historical Evolution and Future Trend

Four major schools of thoughts, primarily from a sociological perspective, contribute to how disasters are understood. These schools focus on social adjustment, human response, technical assessments, and disaster risk management. Sociologists have contributed by (i) laying a foundation for understanding disaster as the product of natural and social forces, (ii) researching how humans respond to disasters, and (iii) linking the risk perception with the concept of disasters in different periods of history. The natural sciences, on the other hand, provide the very basis for disaster mitigation and management approaches through their contributions associated with understanding hazards and risk assessment methods. Researchers from economics, public policy, psychology, and disciplines such as mathematics, toxicology, and public health also study disasters from their perspectives, making the disaster literature vast and rich. Still, many gaps remain in fully understanding the causes, impacts, and ways of effectively responding to disasters.

4.3.1 *Historical Evolution*

The Human Ecology school, with its roots in the disciplines of sociology and geography, has contributed greatly to the understanding of disasters (Mileti and Gailus 2005). Housed within the University of Chicago during the 1900s and influenced by the thinking of John Dewey, the school provided philosophical foundations for geographer Gilbert F. White, who is now known as the father of natural hazard and management research. White was interested in fundamental questions about disasters that are even relevant today: Why are certain adjustments to hazards preferred over others? Why, despite investments in those adjustments, are social losses from hazards increasing? (White 1945; White et al. 1997; Mileti 1999) The basic premise of White's hypothesis is that "natural hazards are the result of interacting natural and social forces and that hazards and their impacts can be reduced through individual and social adjustment" (Mileti 1999, p. 19). This school inspired a range of geographers who study social adjustment processes through land use, planning, protection works, building codes, and building designs.

The second school of thought, which also emerged from the sociology discipline, focused on the question of how humans respond when disaster strikes (Fritz 1961; Drabek 2005). With foundation in the Disaster Research School established by Ohio University in the mid-twentieth century and funded by the National Science Foundation to study population behavior in case of nuclear war, this school focused on post-disaster research—the impact of disasters on people and how people responded to crisis. While subsequent sociologists tried differentiating disaster types and levels (Barton 1969; Britton 1987), a change in thinking occurred when Quarantelli (1987, 1998) suggested moving away from the "agent" or those who caused it. Drabek (2005, p. 7) pressed further to explore such questions as "what are the social processes whereby certain types of crisis situations become 'legitimate' bases for social action?"

Relying on sending research teams in the immediate aftermath of various disasters and comparing one disaster to the other, the school contributed to the theories of social psychology of collective behavior and social organization (Mileti and Gailus 2005). The theories offer explanations for human and organizational adjustment and behavior during and after disasters. Other important findings from the research in this school include the notion that when disaster strikes, emergent networks are born to handle the unique demands generated (Drabek 2005) and that social capital plays a major role in recovering from disasters (Dynes 2002, 2003). Specialized research topics beyond disaster impact studies include warnings, short-term recovery, long-term community reconstruction, and social response to global climate change.

Fairly recently, experts from various fields of the natural sciences such as hydrometeorology, geodynamics, seismicity, mathematics, and toxicology have started undertaking technical assessments of various hazards such as earthquakes, floods, and volcanoes (Cardona 2003). These technical risk assessments take into account past disaster events and project them to the future using trends from the past. Thus, risk became the product of human imaginations disciplined and conditioned by an awareness of the past (Jasonoff 2010). This risk assessment is still a dominant method of calculating disaster risk by generating a list of economic losses that can result from a disaster. The assessments tend to be very "technocratic," focusing on the hazard itself rather than the conditions responsible for it (Cardona 2003). In line with the earlier definition of disaster as an output of natural and social forces (White 1945), risk assessments consider disaster as a product of hazard and vulnerability (Wisner et al. 2004). However, risk assessment still provides the foundation for undertaking disaster mitigation actions.

The current focus on the concept of disaster risk management results from the work of social geographer D. S. Mileti (1999), who sees disaster mitigation as a key part of an environmental agenda, with the core idea of "living with nature and not against it." The approach has been criticized with questions on the meaning of and approach to "sustainability" (Drabek 2005). Table 4.1 provides a summary of the above discussion.

4.3.2 Future Trends: Where Is the Field Heading?

Theories to date, although helpful in understanding disaster situations, fail to explain the increasing monetary losses from disasters. If theories are good and are applied in practice, they should provide a sound basis for reduction in disaster losses and increasing fatalities. To complicate the situation, some mitigation measures only postpone losses that will be more catastrophic. For example, the levee that was protecting New Orleans was only providing a false sense of security until it gave way to a larger catastrophe. Other mitigation measures result in short-term or cumulative environmental degradation and ecological balances (Mileti and Gailus 2005). For example, if land is preserved under flood zoning, some other forest or agricultural

Table 4.1 Contributions of disciplines in current understanding of disasters

	Physical sciences Mathematics Biostatistics Toxicology	Social sciences Human ecology Sociology Economics Public administration	"Sustainable hazard mitigation approach"/ disaster risk management
Key question	What is the probability of failure or accident in mechanical and industrial systems, and what are the likely losses in economic terms?	How do humans respond to disasters? Studies of individuals and their social units, ranging from families to organizations to communities	How to mitigate disasters through an environmentally friendly approach, and how to live with nature and not against it?
Assumptions	Disasters are physical phenomena that generate natural events Losses results not only from the severity of the incident but also from the vulnerability or fragility of exposed elements	Disasters expose the key values and structure that define communities and societies	Actions that are overall better for environment will help in mitigating disaster risk
Major themes in literature	"Focus on hazard" or physical elements exposed to hazards Damages and losses in economic terms	Different definitions of disasters, focus on emergency situations, and vulnerability or risk-based paradigms	Focus on vulnerability Tools of social adjustments
Tools	Risk assessment Risk mitigation	Implementation of a series of strategies and tactics reflecting the full life cycle of disaster, i.e., preparedness, response, recovery, and mitigation	Assess hazard vulnerability; examine possible adjustments; determine the human perception and estimation of the hazard; analyze the decision-making process; and identify the best adjustments, given social constraints, and evaluate their effectiveness

land in a city's periphery will be developed to compensate for the lost land and as population pressure continues.

Although not discussed much in the disaster literature, sociologist Ulrich Beck provides some explanation for the situation. He reminds us that we no longer live in golden era where we can believe that the world works on a rational basis and that we know exactly what is going to happen. In his words:

> How to live in times of uncontainable risks? How to live, when the next terrorist attack is already in our heads? How worried should we be? Where is the line between prudent concern and crippling fear and hysteria? And who defines it? Scientists, whose findings often contradict each other, who change their minds so fundamentally, that what was judged

> 'safe' to swallow today, may be a 'cancer risk' in two years' time? Can we believe the politicians and the mass media, when the former declare there are no risks, while the latter dramatize the risks in order to maintain circulation and viewing figures? (Beck 2006, p. 345)

Beck described a process of "reflexive modernization" in which science not only provides support for the assessment of risk but also skepticism and critique that prevent the construction of stable rationalities to support risk reduction. Based on Beck's thesis, understanding of disasters varies with the concept of risk under three distinct periods: (i) pre-modernity, (ii) industrial society, and (iii) risk society (Jarvis 2007). In the pre-modernity phase, physical risks resulting from hazards such as famines, floods, crop failures, pest infestations, earthquakes, hurricanes, and weather-related hazards were taken as fate or "acts of God," with the attitude that this is how nature is and we need to live with it and that mankind has no or limited power to prevent or change anything. With the advent of the industrial society, people became more cautious about the risk in their life and tried to control it. Risk was measured, assessed, monitored, controlled, and managed. Currently, we seem to be in an era of "radicalized risk" typified by reflexive modernization in which "industrial-technical-scientific projects" produce unintended risks that are incalculable, uninsurable, and beyond the control regime of modernity. Within this landscape, the contribution of earlier schools can be seen during the period of "industrial society" during which the technologies of war were used for peaceful purposes. The guarantee for safety during this period, where risk assessments provided an illusion of control, provided the basis for economic development. The idea was if we can measure it, we can control it. Figure 4.2 below shows the concept of risk, beliefs about disasters, and strategies used for managing or responding to disasters during the three periods.

To a certain extent, the Tohoku earthquake in Japan in 2011 and the following tsunami leading to cascading infrastructure failure including the Fukushima Daiichi nuclear plant crisis have marked a turning point in disaster research. However, even before the great Tohoku earthquake, theorists from public policy and administration had been looking into policies for future events that depend on the degree of certainty or uncertainty. These theorists have focused on the tension between processes of command and control and processes of innovation and discovery to reduce potential collective harm (Comfort 1994b). While strategies to mitigate risk through redundancy and "trial and error" have been discarded due to cost and the potential of catastrophe, respectively, achieving a balance between risk and resilience in complex systems is gaining support.

To cope with the risk of uncertain, destructive collective elements within a complex environment, Wildavsky (1988) suggests a balance between a strategy of anticipation, which is the capacity to prevent harm before it occurs, and a strategy of resilience, which is the capacity to reorganize resources and actions to respond to actual danger after it occurs. Other researchers studying complex system suggest a similar balance between order and chaos (Kauffman 1993) and regularity and randomness (Gell-Mann 1994). Order or regularity is needed to hold and exchange information among component parts, and chaos or randomness is needed for

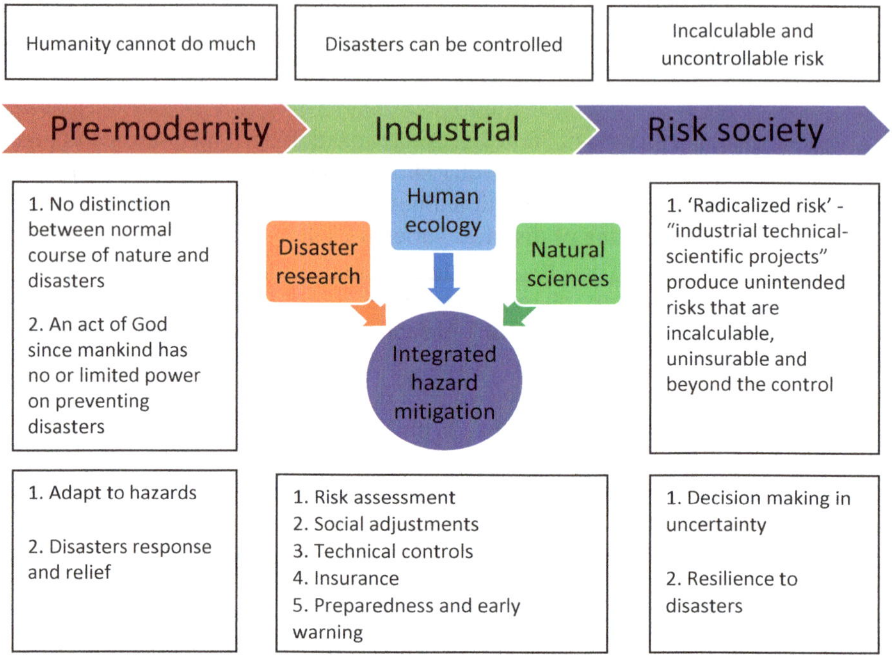

Fig. 4.2 Risk beliefs and response strategies over time (Source: Adapted from Beck (2006) and Jarvis (2007))

innovation and change. Wildavsky (1988) identified four factors to achieve the balance:

- A capacity for creative innovation: Flexibility to adapt to changing situations both internally and externally is likely to generate most innovative strategies in response to unexpected demands.
- Flexibility in relationships between the parts and whole: Risk must be shared throughout the system to improve the safety of its members and the whole system. As parts of the system are interconnected, weakness in one part of the system enables other parts to adjust and lessen weakness throughout the larger system; however, sufficient flexibility is needed for the parts to adjust.
- Interactive exchanges between system and its environment: The process of continuous exchange produces a set of evolving forms resulting in complex, adaptive systems.
- Information exchange: Information flow between constituent parts and the external environment determines its capacity to reduce future risk and create a sustainable relationship with the environment.

Comfort (1994b) argues that disaster response systems should be designed as self-adaptive socio-technical systems for aiding organizational learning and adaptation. Organizational learning takes place through shared mental models developed among members of response operations (Argyris and Schön 1996). However, individuals must learn first for organizational learning to happen (Senge 1990). Organizational learning occurs when members of an organization detect an error and correct it by restructuring the organization's theory of action—called double-loop learning. The change occurs after embedding results of inquiry into its members' minds or in mental models (Argyris and Schön 1996).

Double-loop learning does not occur easily (Senge 1990; Perrow 1999; Birkland 2007) due to the inherent behavior of organizations. Senge (1990) identified four constraints on organizational learning. First, feedback from the real world is often ambiguous, which allows preconceived notions for interpreting available information. Second, post-event adjustment takes place in a highly charged social and political environment which prevents objective analysis of the situation. Third, faulty reporting from individuals with vested interest distorts the reality. Finally, restriction of information flow and secrecy prevents learning.

4.4 Disaster Research Areas

Two key debates emerge from the literature that is related to how the concept of risk is applied. Understanding disasters through the concept of disaster risk presents a very unique kind of framework. Most of the time, we learn about potential damages such as who are at risk and the degree of risk only after risk manifests into a disaster (Jasonoff 2010). Still disaster risk has become a key phrase for understanding disasters and preventing their potential impacts. Technical risk assessment takes into account past disaster events and projects them into the future using trends from the past, much like how population and economic progress are projected into the future. Risk has thus become the product of human imaginations disciplined and conditioned by an awareness of the past (Jasonoff 2010). This risk assessment is still a dominant method of calculating disaster risk and dictates how we assess the economic losses from disasters. The trick is to be able to calculate something that cannot be calculated with certainty. How far can we take the past to guide future trends? How does one know that certain social elements will behave in a certain assumed way? How can one possibly take into account many factors (even with superfast computers) such as climate change, the impact of multiple human actions, and the impact of many unknown factors, which all exist in a dynamic environment?

The "calculus of risk" (Jasonoff 2010) provides an illusion of control and helps administrators in telling the public that all is well—we know what can go wrong and we know what to do when things go wrong. However, it gives a false "perception of risk." Thus, risk maps showing disaster zones may give false information to the people living outside the territory of the "zone," who may be equally at risk compared to those who are in. For example, floods in 2010 in Bihar, India, did not follow the flood risk map as the flood resulted from a dam break that was not considered in the risk assessment. Maps also give wrong information as they do not consider resilience or capacity of people to bear damages (Cardona 2003). Finally, thanks to reliance on risk assessment, current practices of disaster risk management focus largely on physical vulnerability and do not consider social, cultural, economic, and political aspects. No wonder that such an approach does not help in political decisions, making disaster mitigation a low priority (Cardona 2003). Involvement of decision makers, politicians, and communities themselves is thus important in risk assessment. A comprehensive theory of risk is needed that can help in disaster assessment and reduction from multidisciplinary perspectives.

4.4.1 Research in the USA

As noted above, disaster research, especially in the United States, has traditionally focused on three phases of disaster risk management (see Table 4.2 below). Recently, researchers are focusing on 'mitigation and disaster risk reduction' as an ongoing risk management approach rather than distinct phases. Activities and research conducted under the three traditional phases and mitigation and risk reduction approach are discussed below.

4.4.1.1 Preparedness

The pre-disaster phase focuses on anticipating disasters so that resources can be allocated for effective response during disasters. The main instruments used are disaster or contingency plans that spell out the responsibility of all stakeholders—community, individuals, organizations, agencies, and state and national governments—during a disaster. Government machinery remains in charge of testing, updating, and communicating these plans (Mileti and Gailus 2005). Emergency Organization Committees at the local, state, and national levels are also planned beforehand to coordinate actions. Research on this phase focuses on how individuals take decisions for their own safety and what factors contribute to the organizational preparedness. The research conducted so far does not provide a clear picture and is far from comprehensive. For example, researchers know who prepares but not why (Mileti and Fitzpatrick 1993). Preparedness among public sector organizations is better known, but only knowledge regarding organizational preparedness for specific disasters exists. What factors encourage organizational preparedness is also not fully known.

4.4.1.2 Response

This phase involves timely response to a disaster including rescue operations and immediate relief to victims. Actions focus on emergency sheltering, search and rescue, care of the injured, firefighting, damage assessment, and other emergency measures. Increasingly, responders have to focus on coordination, communications, ongoing situation assessment, and resource mobilization during the emergency period, since many emergent groups, volunteers, and NGOs appear after a disaster

Table 4.2 Disaster research focus

	Pre-disaster: preparedness	During disaster: response and relief	Post-disaster: recovery and reconstruction
Aim	Anticipate disasters and prepare for them so resources needed for effective response are in place	Provide timely response to a disaster—relief to victims and restoring situation to normalcy	Putting a disaster-stricken community back together; initial focus on reconstruction of damaged infrastructure and buildings, recently, more focus on decision making in all rebuilding processes
Actions	Formulating, exercising, and testing disaster plans (which shows who will do what during disaster); training responders and general public; developing public awareness campaigns about potential problems during disasters and what to do	Emergency sheltering, search and rescue, care of the injured, firefighting, damage assessment, and other emergency measures: coordination, communications, ongoing situation assessment, and resource mobilization during the emergency period	Usually has three main principles: (1) rapidly return to normal, (2) increase safety, and (3) improve the community
Research areas	Few studies on who prepares for disasters, private sectors, nonemergency-focused agencies and organizations	Most studied phase, focusing on emergent groups and behavior; focus on smaller groups, conceptual frameworks, research designs, and the variables range widely across studies, making generalizations difficult, organizations tend to work in an autonomous and uncoordinated manner	What types of families are most disrupted by a disaster? What family types recover most quickly? What things account for different rates of family recovery? Case studies of specific disasters
Gaps	No thorough understanding of the social-psychological processes involved in making the decision; uneven knowledge about organizational preparedness and the factors that encourage it	Research on how governmental structure and policies influence preparedness and response activities	The processes and outcomes of disaster recovery need to be better understood from a multidisciplinary perspective

Source: Adapted from Drabek (2005), Mileti and Gailus (2005), World Bank (2010)

to help in response activities. Many studies relying on case studies focus on how individuals, families, and groups react during disasters. However, comprehensive studies of how governmental structure and policies influence preparedness and response activities are lacking. While there is relatively little research on state capacity to respond, this research has shown positive change when the response efforts are initiated and managed by lower levels of government. Governmental effectiveness was found to be related to disaster magnitude, the extent to which governmental agencies were prepared, and the public's capacity to cope with disaster impacts (Schneider 1995).

4.4.1.3 Reconstruction

This phase focuses on putting a disaster-stricken community back together (Mileti and Gailus 2005). The immediate impetus of government and affected communities is to focus on the reconstruction of damaged infrastructure and buildings to pre-disaster standards. However, more and more experts are calling for recovery as a process of interaction and decision making among a variety of groups and institutions, including households, organizations, businesses, the broader community, and civil society (Mileti and Gailus 2005). These experts call for focusing on decision-making processes at all levels as they relate to the recovery and risk mitigation process rather than just the physical rebuilding process.

Research on this phase has focused on how reconstruction has taken place after specific disasters, relying primarily on field studies. These studies (Mileti and Gailus 2005) have identified the following factors that hamper disaster mitigation and safer development: complexities of recovery and limited time available, making it difficult to systematically evaluate options; lack of clear goals at federal, state, or local levels; the complexity of acting in concert with multiple entities; and the absence of institutional capacity brought about by advance planning. Success factors include: existence of a preexisting plan or ongoing process for reshaping a community and the availability of outside funding to help bring about the desired changes. Research has also shown that locally based bottom-up recovery efforts are more sustainable in the long run.

4.4.1.4 Mitigation and Risk Reduction

Disaster risk assessments are either used or prescribed for assessing "disaster risk" followed by a range of disaster risk mitigation, reduction, and management options. The idea is to eliminate or reduce the risk wherever possible and transfer those risks that cannot be eliminated or reduced. Various tools are used to transfer and mitigate risks such as: land-use planning and management, building codes and standards, insurance, prediction, forecast and warning, and engineering. In the words of Mileti and Galius (2005), the disaster risk management or integrated hazard sustainability approach:

> calls for increased use of wise, long-term land-use approaches, enhanced production and use of long-term hazard forecasts in community decision-making, insurance as a vehicle to foster mitigation efforts through location decisions and construction practices, and engineered approaches and building codes that go beyond life safety toward protecting the functionality of structures that localities choose to locate in harm's way. (p. 207)

Many researchers have raised questions about the accuracy and comprehensiveness of risk assessments since all risks cannot be predicted due to their very nature as well as whether or not assumptions used in these "technocratic" projections that rely on physical vulnerability and assumptions about human behavior are realistic. Social, economic, and political considerations tend to be missing from such assessments, and thus there remains a gap between "scientific risk assessments" and political decisions to implement their results (Cardona 2003).

An important area of research in the developed countries is to understand the roles of different actors, such as households, groups, organizations, and governments, during different phases of a disaster. The research thus far has focused on how different actors should behave (normative roles) and to some extent on how they are actually behaving (see Table 4.3). Although government roles have traditionally been the subject of this research, recently roles of households and local groups are coming to prominence with the recognition of the importance of social capital for disaster recovery and resilience (Mileti and Gailus 2005). Research on the roles of the private sector and the media is still relatively scarce though the body of research in this area is rapidly developing.

4.4.2 Research in Developing Countries

Current research on disaster risk management in developing countries can be divided into two main areas. The first area, ex ante research, looks at how to improve planning for disaster risk management in developing countries. The second area of research focuses on ex post disaster management, or how people adjust to disasters and governments respond to disasters, including research on organizational learning in the aftermath of disasters. While the ex ante research supports disaster risk management in developing countries with question on "tools and approaches," it tends to be "prescriptive" regarding what can be learned from developed countries and how these "best practices" can be used or grafted onto developing countries. The ex post research area looks at what can be learned from the application of these "relatively new concepts," coming from developed countries, to developing countries. The following main points can be inferred from these studies:

4.4.2.1 Appropriate Policy and Planning Can Reduce Disaster Risk

Land-use planning can reduce disaster risk (Sengezer and Koc 2005). However, in many cities of developing countries, not only is the urban planning field weak but also disaster risk reduction is not mainstreamed into ongoing urban planning practices. International aid organizations can play an important role in integrating urban planning and risk reduction but are not very successful because first, both fields have marginal positions within international aid organizations and, second, there is an incompatibility between the respective professional disciplines (Wamsler 2006).

Table 4.3 Roles of different actors during different phases of DRM

	Normative roles				Actual roles (as shown by researchers)
	Preparedness	Response	Recovery	Mitigation	
Households and families	Have emergency kits, know what to do during emergency	Follow emergency plan	Draw on stocks of social credit, temporary migration	Owning multiple assets, investments to protect and maintain assets, migration, have property and life insurance	Some households prepare and others do not; post-disaster housing pattern reflects pre-disaster social ties, conflict, and socioeconomic status; volunteers during disaster
Groups and nonprofit organizations	Emergency kits, knowing what to do during emergency	Disaster warning	Enhance social capital and help in community recovery	Relocation to safe areas, community training program, community-based information system, and other infrastructure	Not much interest during preparedness except those tasked with crisis response, groups plan in isolation, new emergent groups during response phase
Private sector	Emergency kits, knowing what to do during emergency	Actively participate in emergency plan, prepare one for their business	Existing recovery plan	Private disaster insurance	Not much interest or research
Media	Emergency kits, knowing what to do during emergency	Actively participate in emergency plan, prepare one for their business	Existing recovery plan	Education programs Awareness	As used by the government
Government Local State National	Formulating, exercising, and testing disaster contingency plans, training responders and general public, public awareness campaign about potential problems during disasters and what to do	Emergency sheltering, search and rescue, damage assessment, coordination, communications, resource mobilization	Prepare recovery and reconstruction plan, implement and monitor	Disaster risk profiles, disaster-resilient infrastructure and buildings, adequate social and physical infrastructure	Tasked with emergency response

Source: Adapted from Mileti (1999)

The literature also points to the importance of better urban planning institutions, building regulations, and enforcement in developing countries (Kenny 2009) as many disaster deaths can be avoided with simple engineering solutions, oversight mechanisms, and monitoring.

4.4.2.2 No Standard Tool to Measure Disaster Risk

A number of tools have been proposed to measure disaster risk in urban areas such as hypothetical disaster scenario building, cluster analysis, risk-benefit ratio, risk indexes, and vulnerability scorecard (Hung and Chen 2007; Nicholls et al. 2008; World Wildlife Fund 2009; Munich Re 2006). Most of the available risk assessment tools measure exposure of potential hazards to population and assets. Some tools only assess exposure (to hazards and climate change) in the current situation, while others also consider exposure in future scenarios. The methodologies rarely consider social or economic indicators, although they are important determinants of a community's vulnerability to disasters.

4.4.2.3 Most Government Organizations Do Not Seem to Learn from Past Disasters

A number of studies focus on factors that promote or inhibit organizational learning and adaptation in post-disaster situations with a focus on how disaster response systems evolved over the course of various disasters and what processes facilitate learning and the establishment of an adaptive interorganizational disaster response system. A study conducted in Turkey found that no significant organizational learning occurred within Turkish disaster management following four destructive earthquakes, but the fifth earthquake (Marmara earthquake of 1999) initiated a double-loop learning process that led to changes in the organizational, technical, and cultural aspects of Turkish disaster management (Corbacioglu and Kapucu 2006). Relying on semi-structured interviews of public and nonprofit managers and researchers, the study assessed the changes in organizational, technical, and cultural capacity after each destroying earthquake occurring between 1992 and 1999. The study shows that better information exchange and responses existed during the Düzce earthquake, following organizational changes after the Marmara earthquake. However, it is not clear from the study how much of this change is due to external factors such as timing (because the Düzce earthquake happened just three months after the deadly Marmara earthquake, which caused more than 18,000 deaths compared to less than 800 in earlier quakes), modernization of information technology, and decentralization. The study also failed to show how incremental learning from past disasters contributed to the change after Marmara. It will be interesting to see how much learning has been sustained after the Düzce earthquake.

A similar study conducted in the aftermath of a deadly cyclone in Orissa, India, also pointed to the lack of learning in government organizations (Thomalla and

Schmuck 2004). The affected community, however, learned from prior cyclones and was better prepared.

4.4.2.4 Post-disaster Reconstruction Policy Focusing on Structural Risk Reduction Measures May Increase Disaster Risk

Limited research conducted so far on the impacts of post-disaster reconstruction policy on vulnerability suggests that the pressure to quickly rebuild and focus on reducing structural vulnerability may actually increase the long-term vulnerability of the affected community (Winchester 2000; Ingram et al. 2006). While studying the impacts of disaster prevention strategies in a cyclone-prone area on the east coast of India, the author concluded that the policies to reduce risk in the long term had failed as post-disaster reconstruction benefits were diverted by an alliance of powerful people and recent development reduced the physical protection of the area (Winchester 2000). Similarly, research conducted on the impacts of coastal buffer zone policy adopted by the Sri Lankan government in the aftermath of the 2004 tsunami also suggests that hasty policy to reduce structural vulnerability may actually add to disaster risk rather than reducing it. Instead, vulnerability reduction requires a holistic understanding of the complex interactions between the physical, environmental, and social factors that contribute to it (Ingram et al. 2006). The research in this field relied on qualitative assessment based on interviews and field studies, partly because a credible empirical framework does not exist to evaluate disaster risk. Much of the present understanding of disaster risk is based on the insurance industry's application of probabilistic risk modeling, application of which is difficult in developing countries due to the lack of data and financial resources.

4.5 Toward a Theory for Understanding Disaster Risk Management

In the wake of increasing frequency and magnitude of disaster incidents worldwide, the international community is shifting from a traditionally reactive approach to management of disasters to a proactive or preventive approach. Initially disasters were considered to be one-time events over which mankind had no control. Beginning in the early 1900s, social scientists laid a foundation for understanding disaster as the product of natural and social forces (Cardona 2003; Wisner et al. 2004; Drabek 2005; Mileti and Gailus 2005; World Bank 2010). The idea of "disaster risk" was developed as a product of hazard (Cardona 2003; Drabek 2005), which is a latent natural phenomenon, and vulnerability, which is susceptibility to harm due to the socioeconomic situation (Wisner et al. 2004; McEntire 2004; Enarson et al. 2003). Based on this understanding, capacity to manage "disaster risk" now requires the ability to develop good models of "disaster risk," planning to mitigate known disaster risks, and financially insuring buildings and assets are safe from disasters, along with the ability to be prepared for disaster response such as quick relief, rescue, and rehabilitation operations.

However, even with the new understanding of and increased efforts associated with disaster response and mitigation, monetary losses from disasters are increasing globally. Some researchers point to the failure of disaster risk assessment models, which are not able to predict disasters in time, and mitigation approaches, which are not helpful in reducing higher damages from disasters (Mileti and Gailus 2005). Social scientists now believe that we live in a "risk society," where "industrial-technical-scientific projects" produce unintended risks that are incalculable, uninsurable, and uncontrollable (Beck 2006). To cope, two kinds of abilities are now needed: *anticipation*, the ability to predict and prevent harm before it occurs, and *resilience*, the ability to manage surprise by reorganizing resources and take actions to respond to the actual danger after it occurs (Wildavsky 1988). Resilient and self-adaptive socio-technical systems are needed that can aid organizational learning and adaptation (Comfort 1994a). The capacity to manage disasters now requires an ability to learn, internalize and make change, manage information flow and exchange, ensure flexibility and adaptability in structure and functions of organizations, and coordinate work with multiple agencies and units to achieve a common purpose. However, to date, a comprehensive theory of risk and resilience, especially from a multidisciplinary point of view, is still missing (Cardona 2003) to guide actions on the ground and to answer questions such as what real changes are needed in government structure and functions? Does the community have a role to play? How should the information be managed? How can a "learning culture" be encouraged?

References

Alberti, M., Marzluff, J. M., Shulenberger, E., Bradley, G., Ryan, C., & Zumbrunnen, C. (2003). Integrating humans into ecology: Opportunities and challenges for studying urban ecosystems. *BioScience, 53*(12), 1169–1179.

Anderson, M., & Woodrow, P. (1989). *Rising from the ashes: Development strategies in times of disaster*. Boulder: Westview Press.

Argyris, C., & Schön, D. (1996). *Organizational learning II: Theory, method, and practice*. Reading: Addison-Wesley.

Bankoff, G. (2003). Vulnerability as a measure of change in society. *International Journal of Mass Emergencies and Disasters, 21*(2), 5–30.

Barton, A. H. (1969). *Communities in disaster: A sociological analysis of collective stress situations*. Garden City: Doubleday & Company, Inc.

Beck, U. (2006). Living in the world risk society. *Economy and Society, 35*(3), 329–345.

Birkland, T. (2007). *Lessons of disaster: Policy change after catastrophic events*. Washington, DC: Georgetown University Press.

Britton, N. R. (1987). Toward a reconceptualization of disaster for the enhancement of social preparation. In R. R. Dynes, B. DeMarchi, & C. Pelanda (Eds.), *Sociology of disasters: Contributions of sociology to disaster research*. Milano: Franco Angeli.

Budds, J., Teixeira, P., & SEHAB. (2005). Ensuring the right to the city: pro-poor housing, urban development and tenure legalization in São Paulo, Brazil. *Environment and Urbanization, 17*(1), 89–113.

Cardona, O. D. (2003). The need for rethinking the concepts of vulnerability and risk from a holistic perspective: A necessary review and criticism for effective risk management. In G. Frerks, G. Bankoff, & D. Hilhorst (Eds.), *Mapping vulnerability: Disasters, development, and people*. London: Earthscan.

Comfort, L. K. (1994a). Self-organization in complex systems. *Journal of Public Administration Research and Theory, 4*(3), 393–410.

Comfort, L. K. (1994b). Risk and resilience: Inter-organizational learning following the Northridge earthquake of 17 January 1994. *Journal of Contingencies and Crisis Management, 2*(3), 157–170.

Corbacioglu, S., & Kapucu, N. (2006). Organizational learning and self-adaptation in dynamic disaster environments. *Disasters, 30*(2), 212–233.

Drabek, T. E. (2005). Sociology, disasters and emergency management: History, contributions, and future agenda. In D. A. McEntire (Ed.), *Disciplines, disasters and emergency management: The convergence and divergence of concepts, issues and trends in the research literature*. Emmitsburg: Federal Emergency Management Agency.

Dynes, R. R. (2002). *The importance of social capital in disaster response* (Preliminary Paper #327). Newark: University of Delaware Disaster Research Center.

Dynes, R. R. (2003). Finding order in disorder: Continuities in the 9-11 response. *International Journal of Mass Emergencies and Disasters, 21*(3), 9–23.

Dynes, R. R., Quarantelli, E. L., & Kreps, G. A. (1972). *A perspective on disaster planning*. Columbus: Ohio State University.

Enarson, E., Childers, C., Morrow, B. H., et al. (2003). *A social vulnerability approach to disasters*. Emmitsburg: Federal Emergency Management Agency.

Fritz, C. E. (1961). Disasters. In R. K. Merton & R. A. Nisbet (Eds.), *Contemporary social problems*. New York: Harcourt.

Gell-Mann, M. (1994). *The quark and the jaguar: Adventures in the simple and the complex*. New York: Henry Holt and Company, LLC.

Holling, C. S. (1973). Resilience and stability of ecological systems. *Annual Review of Ecology and Systematics, 4*(1), 1–23.

Holling, C. S. (1986). Resilience of ecosystems; local surprise and global change. In W. C. Clark & R. E. Munn (Eds.), *Sustainable development of the biosphere*. Cambridge: Cambridge University Press.

Hung, H., & Chen, L. (2007). The application of seismic risk-benefit analysis to land use planning in Taipei City. *Disasters, 31*(3), 256–276.

INEE. (2012). *Disaster risk reduction and hyogo framework for action*. http://www.ineesite.org/post/drr/. Accessed on 2 June 2012.

Ingram, J. C., Franko, G., Rumbaitis-del Rio, C., & Khazai, B. (2006). Post-disaster recovery dilemmas: challenges in balancing short-term and long-term needs for vulnerability reduction. *Environmental Science & Policy, 9*(7–8), 607–613.

Jarvis, D. S. L. (2007). Risk, globalization and the nation state: A critical appraisal of Ulrich Beck and the world risk society thesis. *Global Society: Interdisciplinary Journal of International Relations, 21*(1), 23–46.

Jasonoff, S. (2010). A new climate for society. *Theory, Culture and Society, 27*(2–3), 233–253.

Kauffman, S. A. (1993). *The origins of order: Self organization and selection in evolution*. New York: Oxford University Press.

Kenny, C. (2009). *Why do people die in earthquakes? The costs, benefits and institutions of disaster risk reduction in developing countries*. Policy Research Working Paper 4823. Washington, DC: The World Bank.

Lall, S. V., & Deichmann, U. (2009). *Density and disasters: Economics of urban hazard risk*. Development Research Group Policy Research Working Paper 5161. Washington, DC: The World Bank.

McEntire, D. A. (2004). Tenets of vulnerability: An assessment of a fundamental disaster concept. *Journal of Emergency Management, 2*(2), 23–29.

Mileti, D. S. (1999). *Disasters by design: A reassessment of natural hazards in the United States*. Washington, DC: Joseph Henry Press.

Mileti, D. S., & Fitzpatrick, C. (1993). *The great earthquake experiment: Risk communication and public action*. San Francisco: Westview Press.

Mileti, D. S., & Gailus, J. (2005). Sustainable development and hazards mitigation in the United States: Disasters by design revisited. *Mitigation and Adaptation Strategies for Global Change, 10*(3), 491–504.

Munich, R. (2006). *Megacities megarisks: Trends and challenges for insurance and risk management*. Munich: Munich Re Group.

Nicholls, R. J., Hanson, S., & Herweijer, C., et al. (2008). *Ranking port cities with high exposure and vulnerability to climate extremes: Exposure estimates*. OECD Environment Working Papers, No. 1. Paris: OECD Publishing.

Owens, T., Hoddinott, J., & Kinsey, B. (2003). Ex-ante actions and ex-post public responses to drought shocks: Evidence and simulations from Zimbabwe. *World Development, 31*(7), 1239–1255.

Perrow, C. (1999). *Normal accidents: Living with high-risk technologies*. New York: Princeton Paperbacks.

Quarantelli, E. L. (1987). What should we study? Questions and suggestions for researchers about the concept of disasters. *International Journal of Mass Emergencies and Disasters, 5*(1), 7–32.

Quarantelli, E. L. (1998). Introduction: The basic question, its importance, and how it is addressed in this volume. In E. L. Quarantelli (Ed.), *What is a disaster? Perspectives on the question*. London: Routledge.

Schneider, S. K. (1995). *Flirting with disaster: Public management in crisis situations*. Armonk: M. E. Sharpe.

Senge, P. (1990). *The fifth discipline: The art and practice of the learning organization*. New York: Currency Doubleday.

Sengezer, B., & Koc, E. (2005). A critical analysis of earthquakes and urban planning in Turkey. *Disasters, 29*(2), 171–194.

Thomalla, F., & Schmuck, H. (2004). We all knew that a cyclone was coming: Disaster preparedness and the cyclone of 1999 in Orissa, India. *Disasters, 28*(4), 373–387.

UNISDR. (2004). *Living with risk: A global review of disaster reduction initiatives*. Geneva: United Nations International Strategy for Disaster Reduction.

UNISDR. (2009a). *Global assessment report on disaster risk reduction*. Geneva: United Nations.

UNISDR. (2009b). *Terminology on disaster reduction*. Geneva: United Nations.

UNISDR. (2011). *Global assessment report on disaster risk reduction*. Geneva: United Nations.

Vos, F., Rodriguez, J., Below, R., & Guha-Sapir, D. (2010). *Annual disaster statistical review 2009: The numbers and trends*. Brussels: Centre for Research on the Epidemiology of Disasters.

Walker, B., Holling, C. S., Carpenter, S. R., & Kinzig, A. (2004). Resilience, adaptability and transformability in social–ecological systems. *Ecology and Society, 9*(2), 5.

Wamsler, C. (2006). Mainstreaming risk reduction in urban planning and housing: A challenge for international aid organizations. *Disasters, 30*(2), 151–177.

White, G. F. (1945). *Human adjustment to floods*. Department of Geography Research Paper No. 29. Chicago: University of Chicago.

White, G. F., Platt, R. H., & O'Riordan, T. (1997). Classics in human geography revisited: Commentary on human adjustment to floods. *Progress in Human Geography, 21*(2), 423–429.

Wildavsky, A. B. (1988). *Searching for safety*. New Brunswick: Transaction.

Winchester, P. (2000). Cyclone mitigation, resource allocation and post-disaster reconstruction in South India: Lessons from two decades of research. *Disasters, 24*(1), 18–37.

Wisner, B. (2003). Disaster risk reduction in megacities: Making the most of human and social capital. In A. Kreimer, M. Arnold, & A. Carlin (Eds.), *Building safer cities: The future of disaster risk*. Washington, DC: The World Bank.

Wisner, B., Blaikie, P., Cannon, T., & Davis, I. (2004). *At risk: Natural hazards, people's vulnerability, and disaster*. New York: Routledge.

World Bank. (2010). *Natural hazards, unnatural disasters: The economics of effective prevention*. Washington, DC: The World Bank.

World Wildlife Fund. (2009). *Mega-stress for mega-cities*. Gland: WWF International.

Chapter 5
Rethinking the Capacity Development Model

Abstract Researchers have long raised questions regarding the way capacity development is practiced. At the same time, they agree on the need to understand how capacity development takes place before rushing to establish capacity development programs. Many researchers have pointed to the need for understanding the demand side of capacity development, including the need to understand the context, the enabling environment, political will, local governance, and the engagement of civil society organizations. Based on the literature review, a new and holistic conceptual model is presented in this chapter. The Perception Driven Joint Learning Approach (PeDJoLA) model pulls together government's and community's interventions within an environment and disaster risk landscape. Donors and different levels of governments work in a top-down manner, which is matched by bottom-up actions of a community. However, only their combined and integrated efforts can lead to effective capacity development. The chapter starts by providing a brief overview of existing capacity development models and why they provide only a piecemeal understanding of the capacity development process. PeDJoLA is presented with detailed description of its components.

Keywords Change agents • Community capacity • Learning • Local government capacity • Perception Driven Joint Learning Approach (PeDJoLA)

Researchers have long raised questions regarding the way capacity development is practiced; at the same time, they agree on the need to understand how capacity development takes place before rushing to establish capacity development programs. Many researchers have pointed to the need for understanding the demand side of capacity development, including the need to understand the context, the enabling environment, political will, local governance, and the engagement of civil society organizations. A realistic theory of capacity development may identify those factors necessary to understand how endogenous learning and debate relevant to creating context-specific institutions and incremental reform processes occur (Pritchett et al. 2010). Based on the literature review, a new conceptual model is presented in this chapter. The different components of the model are operationalized using concepts and language of the disaster risk management.

This chapter presents a holistic new model for capacity development in disaster risk management. The chapter starts by providing a brief overview of existing

A. Tiwari, *The Capacity Crisis in Disaster Risk Management*, Environmental Hazards, DOI 10.1007/978-3-319-09405-2_5

capacity development models and why they provide only a piecemeal understanding of the capacity development process. A new model of capacity development is presented with detailed description of its components.

5.1 A Conceptual Framework of Capacity Development

Literature on capacity development points to the lack of a universally accepted theory or model that explains how the capacity building process takes place in reality (See Chap. 3). Similarly, literature on disaster risk management points to the lack of a comprehensive theory of risk and resilience, (See Chap. 4) to achieve effective disaster risk management capacity on the ground. This Chapter focuses on filling this gap in the literature, and developing a holistic theoretical model that can increase our understanding of how capacity development in disaster risk management happens on the ground, and how can it be improved. The model will bring together different streams of capacity development models, using concepts and language from disaster risk management.

At the outset, one question is critical to fill the gap in current research on capacity development. What is the actual capacity development process in a specific country or local context? Many capacity development processes start by providing financial resources and developing technical skills. However, mere resources and technical skills are not able to move the capacity development effort to the 'next level'. Brinkerhoff (2008) describes the next level as related to organization, politics and power, and incentives. Although the actual capacity development process is debatable, it is clear that merely providing technical resources and training to develop individual skills is not enough to develop sustainable capacity. Capacity development design needs to consider the environment within which capacity development will take place.

Capacity development and the enabling environment play a mutually reinforcing role. The enabling environment may be assumed to represent a 'bottom-up process,' especially if we include good governance and political will. Thus, we have top-down (donors and state) and bottom-up levels (enabling environment) within which one can assume a capacity development effort should take place. As captured in Fig. 5.1 below, the entire capacity development process can be conceptualized through four main components: (i) top-down planning, (ii) project design, (iii) capacity development, and (iv) bottom-up involvement.

5.1.1 Top-Down Planning

This part of the schematic captures the top-down planning undertaken by the state and the donors (Brinkerhoff 2008; Easterly 2008). Four main criteria seem to drive planning at this level.

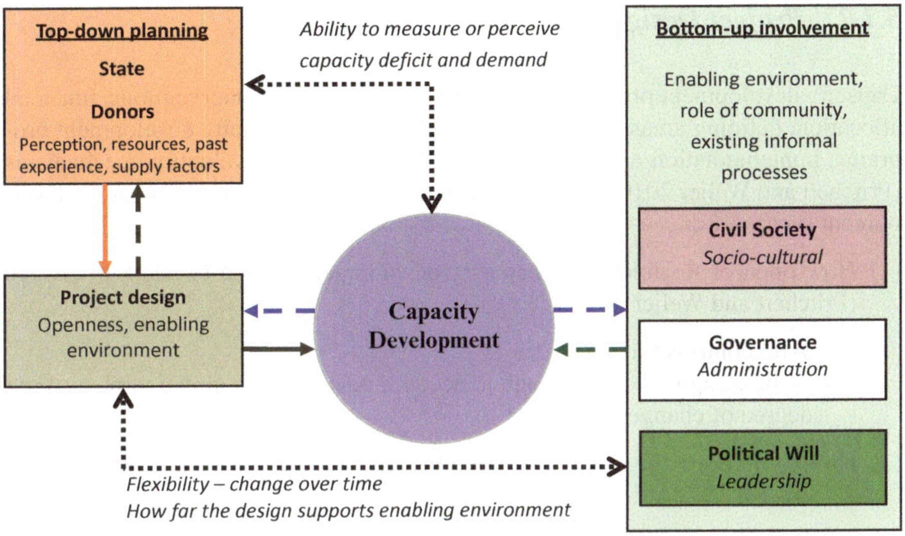

Fig. 5.1 Conceptual capacity development process

1. *The perception of top authorities*: The allocation of resources depends upon how the states and donors perceive the deficit in capacity. The perception of capacity deficit is usually based on those measures, indicators or reports, on which top authorities such as the national and state governments, and donors base their funding decisions. Thus, perception of demand is an important criterion driving top-down planning.
2. *The resources available to allocate and the timeframe within which the resources ought to be used*: Donors and governments have finite resources and a timeline within which they need to plan for capacity development programs and show progress.
3. *The past experience of the government and donors in managing or responding to a situation*: What worked well or did not work in some or other places may also contribute to perceiving the problem or situation in a certain way.
4. *Supply side factors*: The supply side factors are also important, particularly in a donor community when the donor agency has a certain skill set or tool that is to be promoted or is believed to be useful.

Out of these criteria, the ability of donors and national governments to assess the capacity deficit and demand, seems to be crucial for success. National governments already use many statistical measures and assessments to understand capacity deficits and demands. Grindle (2007) has suggested frameworks such as UNDP's Drivers of Change and the World Bank's Country Policy and Institutional Framework to measure capacity deficits.

5.1.2 Project Design

Capacity development project design includes the timing of interventions, financial allocation, defining areas of interventions, objectives of capacity development programs, implementation arrangements, and monitoring and evaluation mechanisms (Pritchett and Weijer 2010). Two important questions with regard to capacity development success are:

(1) How open or flexible is the capacity development design to adapt to change (Pritchett and Weijer 2010)?
   (a) Who controls the design?
   (b) Is the design flexible enough to accommodate evolution in complexity and degree of change over time?
(2) How well does the design support the enabling environment (Brinkerhoff 2010)?
   (a) Does the design consider the enabling environment?
   (b) How did the designer understand the enabling environment and come out with useful suggestions for design?
   (c) How is civil society involved, political will enhanced, and governance improved through the design?

5.1.3 Capacity Development

This component of the conceptual capacity development process shows how capacity is developed in a specific sector of a government's operations. Brinkerhoff (2008) suggested a process, relying on resources (who has what), skills and knowledge (who knows what), organization (who can manage what), politics and power (who can get what), and incentives (who wants to do what); Pritchett and Weijer (2010), suggested an ecological organization and organic growth model that relies on many leaders, internal learning and accountability. The process may also follow a totally different approach, relying heavily on the degree of involvement of local actors and a bottom-up process. The ultimate result of the capacity development intervention is expected to lead to a sustainable capacity to achieve desired goals. For example, capacity development in disaster management should not only bring changes in how national institutions respond to or manage disasters, but should also bring behavioral changes in how people accept the need for disaster prevention. In most cases, this will mean that the project objectives are achieved, and are maintained with innovation in the long run.

Some other elements that are important while thinking about capacity building are (1) the process (Grindle 2007)—how does capacity building take place on the ground, from agenda setting to design, adoption, implementation, and sustainability; and (2) how are the outcomes defined—how do we know whether the capacity development project was successful. Usually, if the project meets its objectives, which has measurable indicators, the project is considered successful by donors and governments. However, these measureable indicators may not show any realistic capacity development on the ground.

5.1.4 Bottom-Up Involvement

The demand side and contextual factors such as civil society organizations showing socio-cultural and bottom-up processes, (good) governance reflecting administration (Grindle 2007), political will and the role of leadership and power (Shiffman and Smith 2007) are an important part of the capacity development process. The following questions are important:

- What is the enabling environment, and how is that supported through the capacity development project (Brinkerhoff 2010)?
- What is the role of the community in capacity development? What is done to sustain momentum?
- Are there existing informal processes? Does the capacity development process affect them?

5.2 Who's and What Capacity?

Two critical questions remain to be answered in the above model: Whose capacity need to be developed, and what kind of capacity is needed? Since all capacity development programs are ultimately implemented at the local level and are meant to influence the local community, capacity development actions ultimately need to target the local level. From disaster risk management perspective, disasters occur first at the local level, and local people respond and experience disasters first hand (Cutter et al. 2012). However, different connotations of local exist, which can range from treating local as community, city, province or a region to watershed regions, ecological zones, or economic regions (Cutter et al. 2012). The larger global context within which local is situated also matters since larger policy, and resource exploitation challenges affects outcome. Thus, linkages within between local, national and regional also need to be considered.

Coming back to the new model, the focus of the model is at local level: (i) local government, (ii) the local communities, (iii) local leaders or agents of change, and (iv) the environment in which capacity development takes place (see Fig. 5.2). The environment needs to capture the global context within which local environment is located, linkages with the higher levels, as well as the local context.

Local government capacity: Polidano (1999) defined public sector capacity with three important elements: *policy capacity*—the ability to structure the decision-

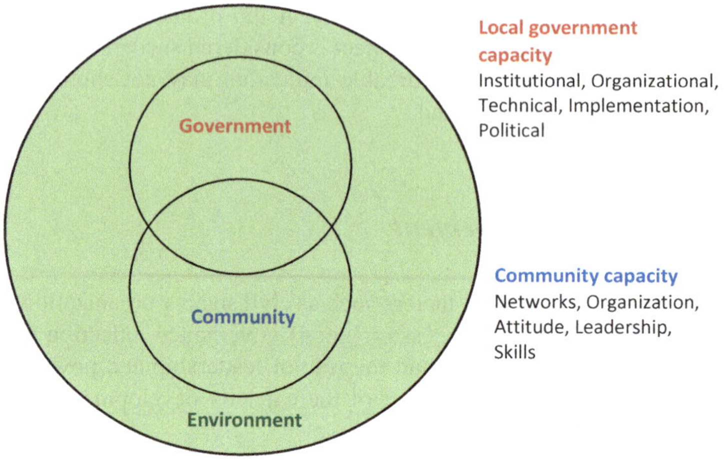

Fig. 5.2 Elements of local capacity

making process, coordinate it throughout government, feed informed analysis into it, and ensure that the analysis is taken seriously; *implementation capacity*—the ability to carry out decisions and enforce rules, within the public sector itself as well as in the wider society; and *operational efficiency* —referring to cost-effectiveness of the internal operations of the public sector and the quality of the services it provides to the public. For the purpose of capturing the government's role in community capacity building, one more element may be important: *responsiveness to the community*—partner and facilitator, how well the government plays the role of and is able to hear local people's concerns and enhance community networks, partnership and capacity (Cavaye 2000).

Grindle (1996) has defined government capacity as consisting of institutional capacity, organizational capacity, implementation capacity, technical capacity, and political capacity. These five dimensions are undertaken in the model.

- *Policy (or institutional) capacity*: the ability to uphold authoritative and effective rules of the game, which comes from enabling laws, policies, and programs.
- *Organizational capacity*: government's internal organization and management style related to structure and distribution of functions; planning; decision making; control and evaluation functions; and information gathering, processing, and distribution.
- *Implementation capacity*: the ability to carry out decisions and enforce rules, within the public sector itself and the wider society.
- *Technical capacity*: the quality of staff, deriving from their skills, knowledge and experience; how are they being used; and whether enough technological, financial and motivational support is available for them to perform effectively.
- *Political capacity*: the ability to mediate conflicts, respond to citizen demands, allow for representation of interests and provide opportunities for effective community and political participation (Grindle 1996).

(i) *Community capacity*: The term community is used in the literature to describe overlapping social units that act as a 'focus of social activity' (Dynes 1998). Community capacity consists of the networks, organization, attitudes, leadership and skills that allow communities to manage change and sustain community-led development (Cavaye 2000). NGOs, governments or the local community itself can organize and enhance community capacity. Government can create local capacity by (Cavaye 2000):

- Creating a vehicle for local people to express and act on existing concerns,
- Judging appropriate interaction with communities from consultation to genuine partnership and facilitation,
- Developing personal relationships between local public servants and community members, which are crucial to the invitation government can receive from local people, and the role government can have in community capacity,
- Melding formal structures that mediate community involvement with a grassroots culture of local participation.
- Helping community members unlearn the idea that the government is simply a provider.
- Unlearning the historical technical assistance approach to communities.

(ii) *Agents or leaders of change*: Leaders such as those existing in national, state, and local governments; non-profit and non-government organizations; donor organizations; and community organizations. These individuals as well as government staff lead the process of transformational change. They work within an environment with multiple ongoing tensions–imperatives and incentives that characterize the space and that either reward or inhibit innovation (Pritchett and Weijer 2010). Their activities jointly can range between the extremes of capacity development and capability trap.

Agents/leaders work within an area shown as the interface between government and community capacity (see Fig. 5.2). This is the area that is important for government and community partnerships to strengthen in order to implement and maintain change. A good example is the development planning process. An urban master plan prepared with the community's active participation stands a better chance of being implemented than a plan prepared without community participation. The extent of involvement can be considered to be manifested in two ways: physical factors such as urban spatial structure (land use and building codes enforcement) and non-physical factors such as a community's disaster preparedness (knowing how to behave during a disaster and having quick access to warning, evacuation procedures, shelters and emergency funds). It is worth noting that the schematic is based on a democratic society since command and control regimes may order the implementation of a development plan and require community preparedness for disasters; however, it is not likely that such one-way enforcement is sustainable in the long run.

Government's responsiveness and civic involvement in government both reinforce each other. Putnam (1993) found that civic involvement improves government in two ways. First, the extent to which people take an interest in public affairs is positively correlated with the performance and efficiency of regional government. Second, public involvement in social groups of any kind, even football clubs and choral societies, generates what Putnam calls social capital: a willingness to submit oneself to rules drawn up in the collective interest. This is then reflected in greater observance of public rules and laws.

(iii) *Environment*: Both government and community capacity exist within an enabling (or disabling) environment. Many researchers have pointed to the role of an enabling environment in capacity and performance. The World Bank's social development sector defines enabling environment on its website as:

> a set of conditions – often inter-related – that impact on the capacity of citizens and civil society organizations to engage in development processes in a sustained and effective manner, whether at the policy, program or project level. They include legal, regulatory and policy frameworks, and political, socio-cultural and economic factors (World Bank 2011).

Polidano (1999) identified five factors affecting the enabling environment: ethnic fragmentation, civil society, political instability, economic crisis and aid dependency. Brinkerhoff (2004) identified five categories of environmental factors as key features of an enabling environment: economic, political, administrative, socio-cultural, and resources. Unless the capacity development process supports the enabling environment, sustainable capacity development cannot be achieved.

5.3 Capacity Development Model as It Applies to Disaster Risk Management

The next step is to operationalize various elements of the conceptual capacity development framework through the language and concepts of disaster risk management. However, before applying the framework to the disaster risk management field, it is important to revisit two capacity development elements that are unique to the disaster literature and need to be introduced into the conceptual framework. They are risk and resilience.

Parallel to the continuum from capability trap to capacity development is the continuum defined by the risk of disasters and resilience. At one end of the spectrum are challenges related to continued risk of disasters. This risk can be either intensive or extensive (UNISDR 2011). *Everyday risks* are related to food insecurity, diseases, crimes, accidents, pollution, lack of sanitation and clean water. *Intensive risks* can result from the exposure of highly concentrated populations and assets to severe intensity hazards. Between these two extremes, there is *extensive* risk, which can result from the exposure of less concentrated populations and assets to lesser or moderate intensity hazards. The continuum of risk can translate into various scales

Fig. 5.3 Continuum of capability trap–capacity development and of risk–resilience

of disasters—from everyday events that occur very frequently but may kill or injure few people, to large disaster events that occur infrequently but often kill or injure large numbers of people and affect an entire city and areas beyond it. Extensive risk can translate into small or medium size disasters that kill a very small number of people and affect localized areas of a city.

At the other end of the spectrum is resilience to disasters. Resilience describes the ability of a system to withstand or accommodate stresses and shocks such as disasters and climate impacts, while still maintaining its function (Comfort 1994b). Disaster resilience depends upon the ability to absorb shocks, to maintain essential assets and to ensure access to services and functions that support the well-being of citizens. Figure 5.3 shows these two parallel continua.

5.4 PeDJoLA: A Simplified Model of Capacity Development in Disaster Risk Management

Building on capacity development models (Figs. 5.1 and 5.2), Fig. 5.4 shows revised model of capacity development in disaster risk management. The model, Perception Driven Joint Learning Approach (PeDJoLA), has six main elements that explain how the capacity development process takes place, and shows a dynamic process based on the perception of State (on capacity deficit) on one side (top down) and communities on the other side (bottom up). Based on perception, State and communities take actions (or project interventions) to develop capacity of governments and communities respectively. Agents, environment, and system play a key role in shaping the capacity development process, which can result into either capacity development and resilience or capability trap and continued risk. The model elements are detailed below (also see Table 5.1).

5.4.1 Perception of Risk and Capacity

This part of the model looks at the perception of disaster risk and their current capacity by both the government and community. Different groups perceive risk and capacity differently. Current psychological research suggests two major generalizations. First, disaster risk is subjective. It involves the probability or

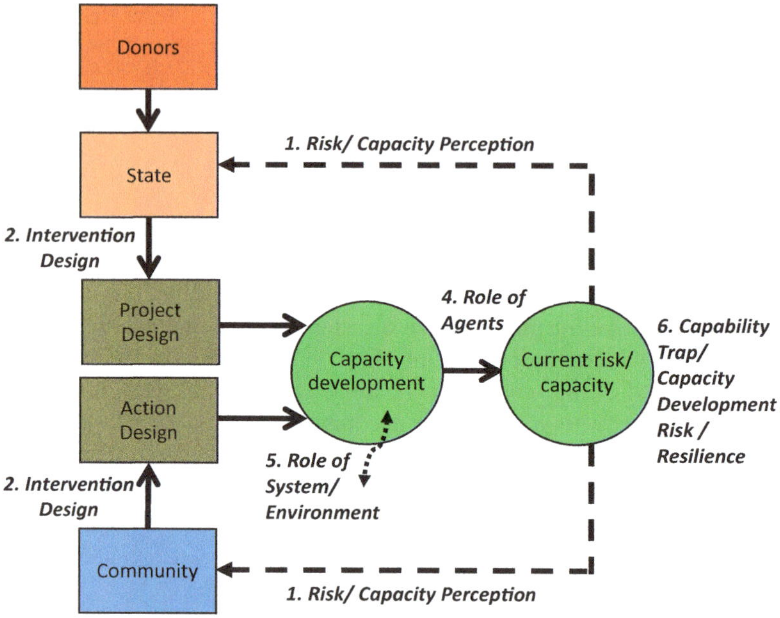

Fig. 5.4 PeDJoLA capacity development model

chance that something bad will happen, and risk is always subjective; what I perceive as bad may not be the same as what others perceive as bad (World Bank 2010). Second, an underlying hypothesis is that people's tolerance for risk is related to their perception of the benefit of living with the risk (Slovic et al. 1982; World Bank 2011). Current research is thin on how communities use their perception to plan for capacity development efforts. Current risk perception research is more geared towards designing appropriate communication strategies to inform the public and understand their behavior during emergencies and stressful situations, and less on understanding how risk perception translates into capacity development efforts.

5.4.2 *Intervention Design*

This part of the model links disasters risk, current capacity, and capacity deficit perceptions with capacity development actions. Based on the assessment of disaster risk in a targeted sector or area, existing capacity, and capacity deficit (comparing what *should be* there to what *is* there), government and donors propose capacity building measures. At the same time, based on the perception of their own risk and capacity deficit in meeting acceptable level of risk, communities take capacity building measures.

Table 5.1 Components and sub-components of the PeDJoLA Model

Capacity components	Sub-components	Application in disaster risk management field
Government capacity		
Institutional	Enabling policies	DRM laws, DRM programs and projects at national, state and local levels
Organizational	Administrative structures	Clear roles and responsibilities in DRM Emergency planning
		Risk mapping and mitigation planning, Monitoring, DRM data collection
	Coordination	Before, during, and after disasters
		Coordination plans or protocols between departments and agencies, Coordination plans
	Publicity	DRM plan publicly available, meetings, periodic DRM material made available
	Awareness	Public awareness plan
Implementation	Enforcement	Building codes, master plan, risk reduction, emergency plan implementation
Technical	Skills and resources	Trained staff, access and motivation for training
	Experience	In handling emergency
Political	Leadership	Awareness and interest in DRM
	Publicity	Public campaigns on DRM
	Community participation	Community involvement in prevention and response planning
	Citizen committees	On DRM and response planning
Community capacity	Skills	On safer house, knowledge of disasters
	Coordination	With different community groups
	Cooperation	Trust and help during disasters
	Leadership	Many community leaders
	Inclusion	Minority, women, poor

5.4.3 Capacity Development Implementation

This includes developing community and government capacity. These terms are described below.

Government capacity: The ability of the local government and organizations to perform successfully and achieve disaster preparedness and resilience. Generally, following ministries deal with areas such as the environment, agriculture, construction, planning, and local government:

- National offices and ministries responsible for emergency management, disaster risk management, and urban development.
- State offices responsible for emergency management, industrial parks, parks and recreation, historic conservation offices, and finance departments.

- City offices such as Public Works and Engineering, Housing, Water and Sanitation, Emergency Management, Transportation, Energy, Finance, Health, and Education.

Government capacity on disaster risk management is defined around five dimensions:

- *Institutional capacity*: From disaster risk management perspective, this would mean having disaster risk management policies, laws, and institutions in place.
- *Organizational capacity*:
  - Administrative structure: Having clear roles and responsibilities in disaster risk management
  - Coordination: Coordination plans and protocols identified between different government departments and agencies
  - Publicity: information on disaster risk management policies, plans, responsibilities are publicly available
  - Awareness: Having public awareness strategy and program
- *Implementation capacity*: This includes enforcement capacity for implementing disaster risk management policies
- *Technical capacity*: Staff with degree or training in disaster risk management
- *Political capacity*: the ability to mediate conflicts, responds to citizen demands, allow for representation of interests and provide opportunities for effective community and political participation (Grindle 1996).
  - *Leadership*: Political leaders having awareness and interest in disaster risk management
  - *Publicity*: Ability to undertake public campaigns on disaster risk management
  - *Community participation*: Ability to involve communities in prevention and disaster response planning
  - *Citizen committee*: Existence of citizen groups or committees on disaster prevention and response

Community capacity: The ability of local communities or neighborhoods to manage all aspects of disaster preparedness on their own as well as in coordination with the government. Community can be understood as consisting of following main classes of people:

- The elite class, which includes decision makers, government officials, businessmen, and politicians.
- The middle class, including middle class government officials, national and local education providers and experts, and private sector employees.
- The urban poor, which includes daily wage earners, lower-level government employees and workers in informal sectors.

All community actors are interconnected (see Fig. 5.4) through various formal and informal processes. There are national and state governments; international, national and regional NGOs; and financial institutions, all of which influence the formal and informal decision making at the local level. The middle class and the urban poor must be strengthened and activated to ensure participatory and transparent planning processes. Community capacity consists of skills, coordination, cooperation, leadership and inclusion.

Community capacity with regard to disaster risk management consists of

- *Skills*: Understanding disaster risk management, the risk communities are facing, and their skills such as those related to building safer houses
- *Coordination*: Amongst and between different community groups and with the government
- *Leadership*: Having many community leaders aware of and involved in disaster risk management
- *Inclusion:* Communities include minority, women and poor

5.4.4 Role of Agents

Leaders of change such as government managers at national, state and local levels; non-profit and NGO leaders, representatives of donor organizations; community leaders; political leaders and government staff. They work within an environment with multiple ongoing tensions—imperatives and incentives that characterize the space and that either reward or inhibit innovation (Pritchett and Weijer 2010). Their activities jointly can range from one extreme of capacity development and resilience to the other extreme of capability trap and vulnerability.

5.4.5 System and Environment

Both the government and the community capacity exist within a dynamic system. The system can be measured through the openness, flow of information, and relationships between the parts and the whole. The system exists in balance with its environment. The environment consists of economic, political, administrative, and socio-cultural characteristics (Brinkerhoff 2004).

Since disaster events and the risk they pose are uncertain, disaster response systems need to be self-adaptive socio-technical systems for aiding organizational learning and adaptation (Comfort 1994a). Whereas the capacity development literature treats the enabling environment as a rather passive element affecting the capacity development process, a system within which disaster risk management efforts are undertaken needs to be 'active' and 'dynamic.' This can ensure an adaptable and efficient system that is able to manage surprises and bounce back after a calamity. Four factors are important for such a system (Wildavsky 1988):

- Flexibility to adapt to changing situations: This is likely to generate innovative strategies in response to unexpected demands.

- Flexibility in relationships between the parts and the whole: Risk must be shared throughout the system to improve safety of its members and the whole system. As parts of the system are interconnected, weakness in one part of the system requires other parts to adjust and reduce weaknesses throughout the larger system. However, the parts require sufficient flexibility to make these adjustments.
- Interactive exchanges between system and its environment: The process of continuous exchange produces a set of evolving forms resulting in complex adaptive systems.
- Information exchange: Information flow between constituent parts and the external environment determines the system's capacity to reduce future risk and create sustainable relationships with elements in the environment.

5.5 Capability Trap or Capacity Development Risk/ Resilience

Capacity development in disaster prevention and response is a locally-driven process of transformational learning by local organizations, community, leaders, and other agents that leads to disaster resilience. Capacity development will result from a system that is open, flexible, and adaptable to change; relying on functionality, innovation, empowerment, inclusiveness and demonstrated success; and with agents leading the learning process, contributing to public value creation, and bringing desired change by altering the system and the environment. As the capacity develops, the communities will have an enhanced ability to anticipate disasters and readiness to deal with disasters before, during, and after disasters occur through risk assessments, risk reduction plans and programs, contingency plans, emergency management structure and resources, clear lines of communications, and awareness. This will reduce disaster impacts and will help the community get back to normalcy relatively quickly after a disaster.

A capability trap will result in an inability to achieve performance or desired collective goals for a long time, even after implementing conscious capacity development efforts. Such inability will be related to weaknesses along one or more dimensions of local government and community capacity. It results from a closed system that does not provide flexibility and adaptability, organizational factors such as reliance on the form rather than on function, and pressure to perform more than the existing capacity, which will ultimately leading to collapse. At the level of agents, elite capture[1] and rent seeking[2] activities will lead to personal gain but no public value creation or transformational learning to drive change. Capability traps

[1] Elite capture is where public resources are appropriated by a few individuals of superior social/ economic/ political status.

[2] Rent seeking is where individuals or groups spend resources in order to increase one's share of existing wealth, instead of trying to create wealth.

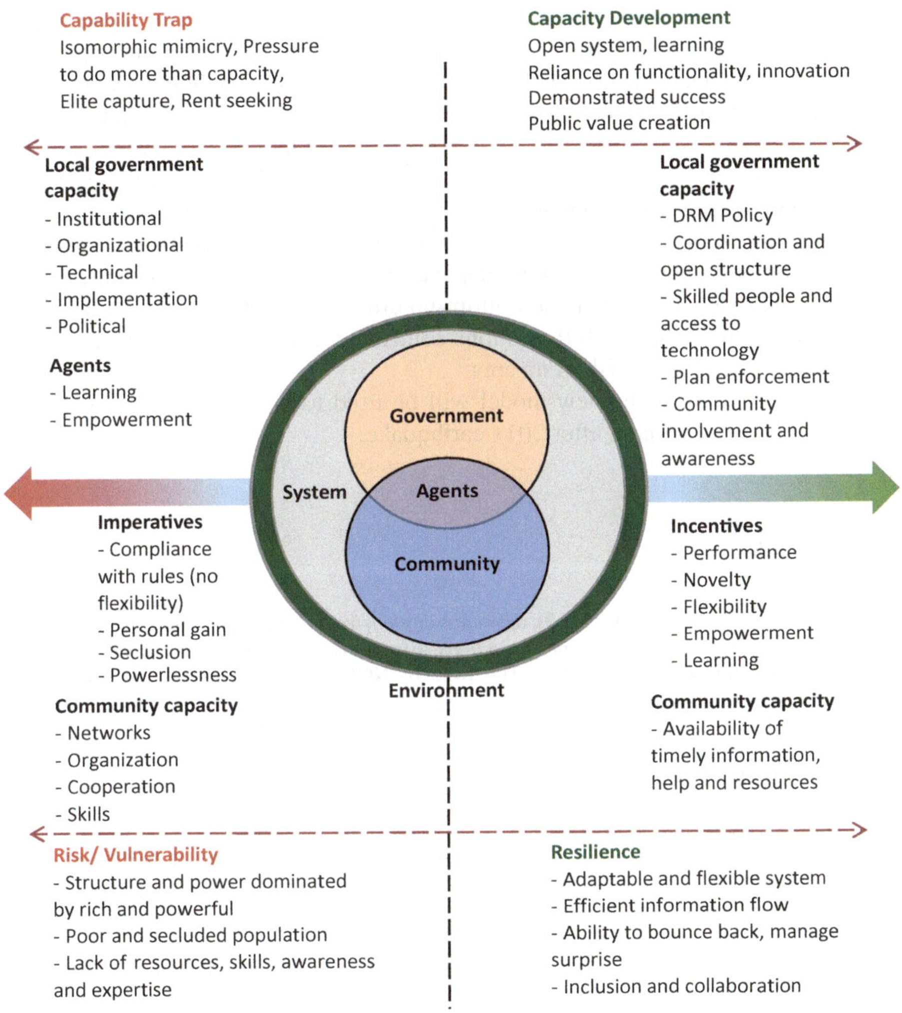

Fig. 5.5 Capacity development in disaster risk management

will lead to situations where vulnerability to disasters remains high. In such situations, disaster impacts will keep increasing.

Components of the PeDJoLA conceptual framework as they apply to the DRM field are further detailed in Fig. 5.5 and Table 5.1.

This chapter presented a holistic conceptual model, which pulls together different concepts and theories from the capacity development and disaster risk management fields. The model brings together top-down and bottom-up interventions within an environment and disaster risk landscape. Donors and different levels of governments work in a top-down manner, which is then matched by bottom-up actions of a community; however, only their combined and integrated efforts can lead to effective capacity development. The perception of current capacity and disaster risk plays a key role in determining actions, which are implemented primarily by 'change agents' such as government officers, political leaders, and community leaders at the local level. While change agents work within a landscape of incentives and imperatives, the entire capacity building process works along a continuum. At one extreme, the combined efforts can lead to sustainable capacity building and disaster risk reduction and, on the other extreme; the efforts can lead to a capability trap or continued vulnerability to disasters. Momentum for sustaining capacity building efforts not only depend upon the motivation of change agents and how integrated top- down and bottom-up efforts are, but also on how information is shared, and how much the actions contribute to improving the environment and reducing the disaster risk of an area.

In the next chapter, the new model will be used to better understand capacity building process in Gujarat after 2011 earthquake.

References

Brinkerhoff, D. W. (2004, May 12–13). *The enabling environment for implementing the millennium development goals: Government actions to support NGOs*. Paper presented at George Washington University conference "The Role of NGOs in Implementing the Millennium Development Goals", Washington, DC.

Brinkerhoff, D. W. (2008). The state and international development management: Shifting tides, changing boundaries, and future directions. *Public Administration Review, 68*(6), 985–1002.

Brinkerhoff, D. W. (2010). Developing capacity in fragile states. *Public Administration and Development, 30*(1), 66–78.

Cavaye, J. M. (2000). *The role of government in community capacity building*. Department of Primary Industries and Fisheries Information Series QI99804, Toowoomba, Queensland: Queensland Government.

Comfort, L. K. (1994a). Self-organization in complex systems. *Journal of Public Administration Research and Theory, 4*(3), 393–410.

Comfort, L. K. (1994b). Risk and resilience: Inter-organizational learning following the Northridge earthquake of 17 January 1994. *Journal of Contingencies and Crisis Management, 2*(3), 157–170.

Cutter, S., Osman-Elasha, B., & Campbell, J., et al. (2012). Managing the risks from climate extremes at the local level. In C. B. Field, V. Barros, T. F. Stocker, D. Qin, D. J. Dokken, K. L. Ebi, M. D. Mastrandrea, K. J. Mach, G.-K. Plattner, S. K. Allen, M. Tignor, & P. M. Midgley (Eds.), *Managing the risks of extreme events and disasters to advance climate change adaptation. A special report of working groups I and II of the Intergovernmental Panel on Climate Change* (pp. 291–338, 582 p.). Cambridge/New York: Cambridge University Press.

Dynes, R. R. (1998). Coming to terms with community disaster. In E. L. Quarantelli (Ed.), *What is a disaster?* New York: Routledge.

Easterly, W. (2008). Institutions: Top down or bottom up? *American Economic Review, 98*(2), 95–99.

Grindle, M. (1996). *Challenging the state: Crisis and innovation in Latin America and Africa*. Cambridge, MA: Cambridge University Press.

Grindle, M. S. (2007). Good enough governance revisited. *Development Policy Review, 25*(5), 553–574.

Polidano, C. (1999). *Measuring public sector capacity*. IDPM public policy and management working paper. Manchester: University of Manchester.

Pritchett, L., & Weijer, F. (2010). *Fragile states: Stuck in a capability trap?* Input paper for The World Development Report 2011. Washington, DC: The World Bank.

Pritchett, L., Woolcock, M., & Andrews, M. (2010). *Capability traps: The mechanisms of persistent implementation failure*. Working Paper 234. Washington, DC: Center for Global Development.

Putnam, R. D. (1993). *Making democracy work: Civic traditions in modern Italy*. Princeton: Princeton University Press.

Shiffman, J., & Smith, S. (2007). Generation of political priority for global health initiatives: A framework and case study of maternal mortality. *The Lancet, 370*(9595), 1370–1379.

Slovic, P., Fischhoff, B., & Lichenstein, S. (1982). Why study risk perception? *Risk Analysis, 2*(2), 83–93.

UNISDR. (2011). *Global assessment report on disaster risk reduction*. Geneva: United Nations.

Wildavsky, A. B. (1988). *Searching for safety*. New Brunswick: Transaction.

World Bank. (2010). *Understanding risk: Innovation in disaster risk assessment*. Washington DC: The World Bank. Available at http://www.understandrisk.org/ur/sites/default/files/attachments/Forum2010.pdf. Accessed on 15 Nov 2011.

World Bank. (2011). *Enabling environment for civic engagement*. Available at http://go.worldbank.org/4JJJ1MKSZ0. Accessed on 12 May 2011.

Chapter 6
Revisiting Gujarat: Is There a Capability Trap?

Abstract This chapter applies the Perception Driven Joint Learning Approach (PeDJoLA) model of capacity development to understand Gujarat's post-2011 earthquake capacity building process. The chapter focuses on understanding capacity building efforts from two levels: top-down level, within different levels of governments, and bottom-up level, within different community groups. Sustainability of capacity development efforts is also discussed along with challenges that inhibit developing effective capacity. The chapter presents findings from field research carried out in 2012, through interviews, surveys, and review of documents. A methodology to apply the conceptual model through field research is discussed first. The second section describes the perception of research participants regarding capacity deficit and disaster risk. The third section describes different forms of capacity development efforts undertaken by different stakeholders, including the government and the community. The fourth section looks into whether capacity development under the Gujarat Emergency Earthquake Reconstruction Project (GEERP) has taken place or not, focusing on changes in the local government and the community. The fifth section explores how sustainable the current capacity development efforts are, in terms of learning and empowerment. The final section, which relies heavily on formal and informal interviews, highlights the potential factors responsible for inhibiting capacity development in the three study towns.

Keywords Capacity deficit • Perception • Inclusion • Sustainability • Gujarat Emergency Earthquake Reconstruction Project (GEERP)

Is Gujarat facing a capability trap? In response to this critical question raised in Chaps. 2, 3, 4, and 5 focused on understanding what capacity building and disaster risk management means, and how capacity building in disaster risk management takes place. In this chapter, I apply the PeDJoLA capacity development model to understand Gujarat's post-2011 earthquake capacity building process. The chapter focuses on understanding capacity building efforts from two levels, at top-down level, different levels of governments, and bottom-up level, different community actors. Sustainability of capacity development efforts is also discussed along with challenges that inhibit developing effective capacity. The chapter presents findings from field research carried out in 2012, through interviews, surveys, and review of documents.

A. Tiwari, *The Capacity Crisis in Disaster Risk Management*, Environmental Hazards, DOI 10.1007/978-3-319-09405-2_6

A methodology to apply the conceptual model through field research is discussed first. The findings from the field research are then presented in relation to the PeDJoLA model, beginning with a brief overview of the environment and the disaster risk of the study area. This includes the potential hazards to which the study area is exposed and the socioeconomic factors within which the capacity building process under the Gujarat Emergency Earthquake Reconstruction Project (GEERP) is taking place. Later sections focus on the capacity building process through findings with respect to the five main research questions. These six questions represent the entire conceptual model, from the assessment and perception of current capacity to the designing and implementation of capacity development efforts. The second section describes the perception of responders regarding capacity deficit and disaster risk. The third section describes different forms of capacity development efforts undertaken by different stakeholders, including the government and the community. A link between capacity deficit perception and capacity development efforts is also explored. The fourth section looks into whether capacity development under the GEERP has taken place or not, focusing on changes in the local government and the community. The fifth section is about how sustainable the current capacity development efforts are, in terms of learning and empowerment. The final section, which relies heavily on formal and informal interviews, highlights the potential factors responsible for inhibiting capacity development in the three study towns.

6.1 Assessing Capacity Development Under GEERP

The PeDJoLA model was used to assess the entire capacity building process under GEERP. Six key questions were identified relating to six key components of the conceptual model (see Tables 6.1 and 6.2):

1. How do different actors perceive the capacity deficit just after the earthquake and now?
2. What were the different forms of capacity development undertaken in Gujarat after the 2001 earthquake?
3. What factors related to the enabling environment lead to effective and sustainable capacity development?
4. Did the GEERP undertaken by the government of Gujarat result in capacity development?
5. Was the capacity development of local governments and communities, as achieved under GEERP, effective and sustainable?
6. What factors inhibit capacity development or lead to a capability trap?

The key questions were further divided into subcomponents for collecting and analyzing data. These are given in Table 6.2 below.

Table 6.1 PeDJoLA model and key questions

Model components	Key questions (supportive questions are italicized)
1. Risk and capacity perception	How do different actors perceive the capacity deficit and disaster risk just after the 2001 earthquake in Gujarat and now?
	What are the factors contributing to local capacity, specifically related to reducing the impacts of disasters?
2. Intervention design	What were the different forms of capacity development undertaken in Gujarat after the 2001 earthquake?
	Are the adopted capacity development forms based on perceived capacity deficits?
3. Capacity development implementation	What was the process of capacity development implementation?
	Did the GEERP undertaken by the government of Gujarat result in capacity development?
4. Role of agents	What was the role of agents in capacity development implementation?
	What incentives for agents lead to effective and sustainable capacity development?
	What imperatives for agents inhibit capacity development or lead to capability trap?
5. Role of enabling environment	What factors related to the enabling environment lead to effective and sustainable capacity development?
6. Capability trap or development, risk or resilience	Did the GEERP undertaken by the government of Gujarat result in capacity development?
	Was the capacity development of local governments and communities, as achieved under GEERP, effective and sustainable?
	What factors inhibit capacity development or lead to capability trap?

Table 6.2 PeDJoLA model components

Components	Subcomponents
Government capacity	
Institutional: The ability to uphold authoritative and effective rules of the game which comes from enabling law, policies, and programs	*Enabling policy*: DRM law and policy exists within which local governments are given resources and guidance for plans and programs
	Programs and plans on disaster preparedness, risk assessment, alert, micro-zoning and awareness exist and are routinely updated
Organizational: The government's internal organization and management style related to structure and distribution of functions, planning, decision-making and control and evaluation functions, information gathering, processing, and distribution	Administrative structure: Different government agencies have clear roles and responsibilities related to DRM
	Coordination: Agencies coordinate their actions before, during, and after disasters
	Information sharing: Disaster alert and prevention information system flow

(continued)

Table 6.2 (continued)

Components	Subcomponents
Implementation: The ability to carry out decisions and enforce rules, within the public sector itself and the wider society	Enforcement: The ability to implement and monitor building codes, master plan, risk reduction, emergency plan
Technical: The quality of staff, deriving from their skills, knowledge and experience, how are they being used and whether enough technological, financial and motivational support is available for them to perform effectively	Skills: DRM trained staff know what to do before, during, and after disasters in order to minimize disaster impacts
	Experience: of handling emergency through on-the-job training or courses
	Motivation: Financing support for DRM training, performance awards for readiness
	Infrastructure: Such as seismic stations, alert system, information management, risk assessment, and sharing platform to help in being ready for disaster
Political: The ability to mediate conflicts, respond to citizen demands, allow for representation of interests, and provide opportunity for effective community and political participation	Leadership: Many political leaders are giving increasing attention to DRM-related issues
	Publicity: Ability to increase public awareness on DRM
	Community participation: Community involvement in prevention and response planning
	Citizen committees: Committees related to DRM exist and are supported by the government
Community capacity: The capability of local communities or neighborhoods to manage all aspects of disaster preparedness on its own as well as with the government	Skills and resources: have sufficient knowledge and skills for safe home construction and disaster prevention
	Coordination: Between different community groups
	Cooperation: Trust and help during disaster
	Leadership: Existence of many community leaders raising voices on DRM
	Inclusion: Community activities involve minority, women, poor
Learning: The internalization of lessons from past experiences and passing them on to the successors (both in government and community)	Flexibility and adaptability: To take own actions
	Information sharing and flow: The ease of information flow, availability, and dissemination
	Learning culture: Ability to reflect on previous disasters and make change for preventing future impacts personally and as a group
	Innovation: Ability to make changes into routine efforts
	Double-loop learning: Individual, organization, or entity is able, having attempted to achieve a goal on different occasions, to modify the goal in the light of experience or possibly even reject the goal (Argyris and Schön 1996)

(continued)

Table 6.2 (continued)

Components	Subcomponents
Empowerment: The process of enhancing individual or group capacity to make choices and transform those choices into desired actions and outcomes (Gibson and Woolcock 2005). *A group is empowered when disparate members of the group know their rights and choices, express these by democratic means, and bring about the desired result*	Awareness: Choices and rights
	Democratic process: All groups are fairly represented in major decision-making
	Collaboration: Government agencies and communities work together to ensure disaster preparedness
	Ownership: Community maintains and participates in planning and programs
	Minority participation: Minority group participation is encouraged
	Choices available: Many choices are available to communities when planning for disaster preparedness

Box 6.1. Definitions of Key Terms

Capacity development: A locally driven process of transformational learning by local organizations, community, leaders, and other agents that leads to disaster resilience.

Capability trap: The inability to achieve performance or desired collective goals for a long time, even after conscious capacity development efforts. In the context of developing countries, this inability may be coming from continued reliance on donors.

Local government capacity: The ability of the local government and organizations to perform successfully and achieve disaster preparedness and resilience. Local government capacity has five dimensions: policy, organizational, implementation, technical, and political.

Disaster resilience: The state of being able to prevent disaster impacts and bounce back to normalcy after disasters. This includes the ability to anticipate disasters and readiness to deal with disasters before, during, and after they occur through risk assessments, risk reduction plans and programs, contingency plans, emergency management structure and resources, clear lines of communications, and awareness.

Sustainability: The effectiveness and continuity of local capacity already developed as well as subsequent additions, even when there are changes in the personnel (agents).

Learning: The internalization of lessons from past experiences and passing them on to the successors (both in government and community). Learning comprises flexibility, information sharing and flow, "double-loop learning," openness to change, and innovation.

Empowerment: The process of enhancing individual or group capacity to make choices and transform those choices into desired actions and outcomes (Gibson and Woolcock 2005). A group is empowered when disparate members of the group know their rights and choices, express these in democratic means, and bring about the desired result.

6.2 Research Methodology: Focusing on Three Towns in Kutch

Kutch district was the worst-affected by the 2001 earthquake and received much attention from GEERP. Three towns in Kutch district were identified for detailed assessment—Bhuj, Bhachau, and Mandvi. Bhuj and Bhachau received large sums of post-disaster reconstruction funding, have strong NGO presence, and created new agencies (area development authorities) for planned development. Mandvi, the town that was least affected by the 2001 earthquake and received the least attention from government agencies and others, was used to separate out any bias from disaster effects (see Fig. 6.1 for a map of the area).

All the government agencies dealing with disaster management as well as a number of key resource persons (those running citizen groups, retired senior government staff, etc.) were interviewed or surveyed. The major NGOs, CBOs, interest groups, and citizen groups working in these towns were the third section of the study population. Though not all of them deal directly with disaster management, they do deal with community empowerment and capacity building. Many academic institutions and private consultants worked with the communities or carried out research related to capacity development initiatives. They also were useful in obtaining information. Finally, local business owners, volunteers, and political leaders were included to get a full picture of community capacity or resilience. See Appendix for details.

6.2.1 The Role of Environment and Disaster Risk in the Study Area

Based on the conceptual model, the environment and capacity development play a mutually reinforcing role. Thus, the starting point in understanding the capacity development process under the GEERP is to know the "environment" within which capacity development is taking place. To capture this understanding, this section describes the physical and social "environment" within which the capacity building process under the GEERP is taking place. Both these elements also contribute to the disaster risk of the area. Capacity development in disaster risk management takes place within a current landscape of disaster risk, which in turn helps reduce the disaster risk. Figure 6.2 below shows this process.

The environment and disaster risk factors could be understood through two contributing factors: hazards and vulnerability (socioeconomic) factors. Hazards represent latent danger or external risk factors (Cardona 2003; UNISDR 2004, 2009) that represent potential harm to a community or an environment (Drabek 2005). To understand the kind of hazards that the study area is exposed to, the hazard profile of the study area is discussed. Vulnerability is an internal risk factor that includes economic, social, political, and physical susceptibility or predisposition of a

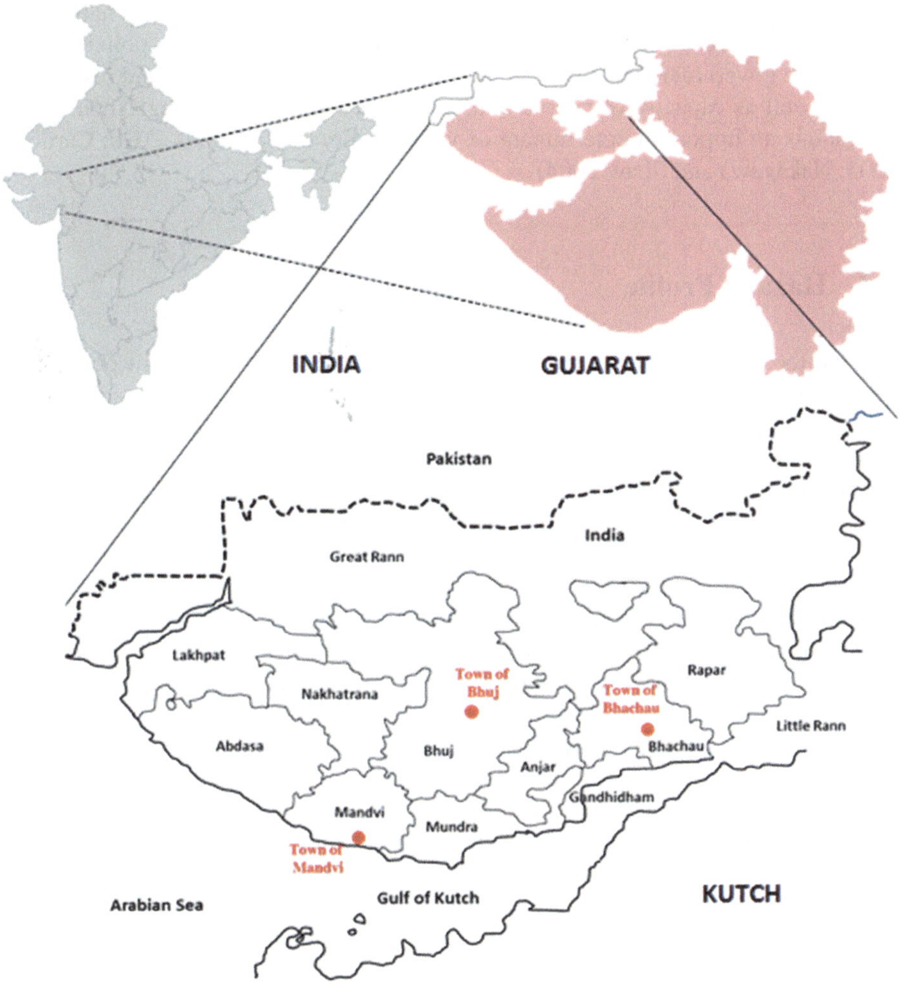

Fig. 6.1 Location of the three study towns

Fig. 6.2 Disaster risk and capacity development

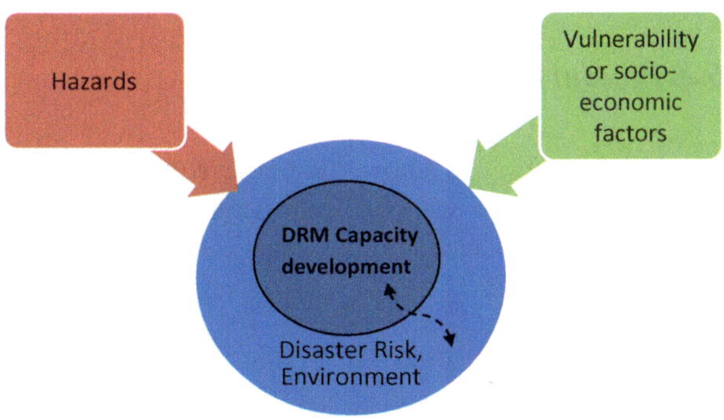

community to damage in case of a destabilizing phenomenon of natural or ethnographic origin (Cardona 2003). The capacity of institutions—policies, programs, and government machinery to prevent disasters, alert people, and respond in time as well as capacity of people and community measured through their social capital—is an important determinant of vulnerability (World Bank 2010; Cardona 2003; Nakagawa and Shaw 2004).

6.3 Hazard Profile

Kutch is the largest district in India in terms of land area but has very low population density since a large part of the district is uninhabitable Rann (shallow salt marshland that is submerged in water during the rainy season and becomes dry salt flats during other seasons). Due to its location and climate, Kutch district is prone to earthquakes, cyclones, floods, and drought. The district is an active seismological area with a number of active faults, primarily running east-west. Kutch falls under seismically active zone V[1] with a history of major and minor earthquakes over the years. The dryness of the area and dependence on Monsoon rains for water keep the area under drought and drought-like conditions for most of the year. Only 15 % of the area is cultivable (Mehta 2001). Average rainfall ranges from 34 to 44 cm (Raju 1995). At the same time, the coastal areas are under constant threat of cyclones and floods.

Kutch has witnessed many destructive earthquakes in the past; well-known among them are 1819 Allah Bund (magnitude 8) and 1956 Anjar (magnitude 7). In the span of just 50 years, the Kutch region has experienced two large magnitude earthquakes: the July 21, 1956, Anjar (magnitude 7) and the 2001 event (magnitude 6.9). The 2001 earthquake resulted in around 12,000 deaths and destroyed nearly 250,000 houses in the Kutch district alone (Government of Gujarat 2001). New studies are suggesting that the existing faults still have high potential for a large magnitude earthquake in the future. Although it is still impossible to predict earthquakes, geologists believe that magnitude 8 earthquakes are very likely in the Kutch region (Sato et al. 2001). See Table 2.2 for major disasters in Gujarat.

6.3.1 *Earthquake Hazard*

As shown in Fig. 6.3 below, the entire Kutch district is earthquake prone. The talukas (administrative unit) of Bhuj, Anjar, and Bhachau face the most intense hazard level (the darker shades on the maps).

[1] Where earthquakes of magnitude 8 on Richter scale can be expected.

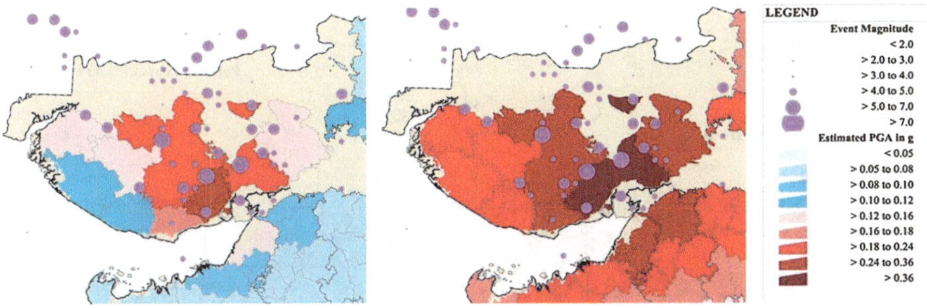

Fig. 6.3 Earthquake hazard risk zonation (50 and 100 year return period) (Source: GSDMA (2012))

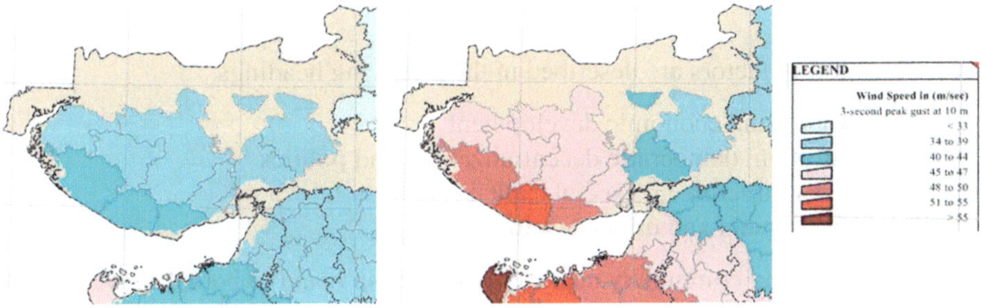

Fig. 6.4 Cyclone hazard risk zonation (50 and 100 year return period) (Source: GSDMA (2012))

6.3.2 Cyclone Hazard

As shown in Fig. 6.4 above, the entire Kutch district is cyclone prone. The talukas of Mundra, Mandvi, and Abdasa face the most intense hazard level (the darker shades on the maps).

6.3.3 Flood Hazard

As shown in Fig. 6.5 below, there are some pockets in the district where flood becomes an issue during heavy rains. These are at and near small dams across seasonal water channels. However, during my field visit, local people pointed out that localized floods due to blockage of natural water channels in the urban areas are more of a concern for the city residents than these dam overflows.

From the hazard profile presented above, it is clear that the entire district is at a severe risk of multiple hazards.

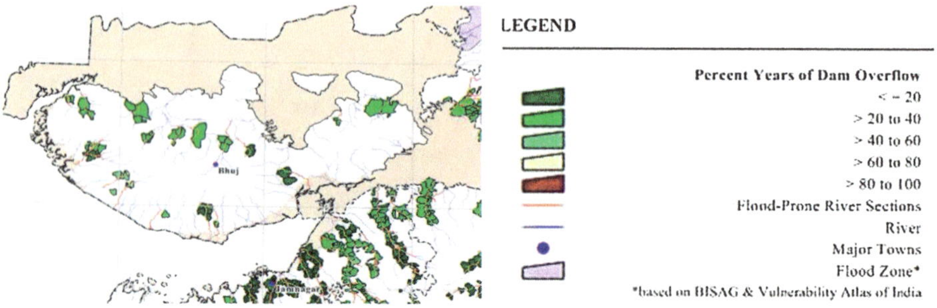

Fig. 6.5 Flood hazard risk zonation (Source: GSDMA (2012))

6.4 Vulnerability or Socioeconomic Factors

The vulnerability factors are described in the following headings:

(a) Demography and economic development
(b) Administration, democratic decentralization, and politics
(c) Community structure and social capital
(d) Built environment and urban development

Government and community capacity (captured in b and c) both deserve special attention as these also represent the bottom-up and demand side of capacity development.

6.4.1 Demography and Economic Development

The population of Kutch recorded in the 2011 census was close to two million (see Table 6.3). Compared to national and state averages, the district has a very low density (46 persons per square kilometer), which reflects the large area covered by seasonal wetlands. About 35 % of district's population is urban. The literacy rate is higher than the national average but less than the state average. Similar to national average, there is a wide difference in the literacy rate of men and women in Kutch. While 80 % of men are literate, only 61 % of women are literate as per 2011 census.

Based on the population, the urban areas are classified into municipal corporations or municipalities (Classes A to D). The demographic details for the three towns studied are presented in Table 6.4 below. Not only is Bhuj the largest town in the district, it has a large percentage of its population living in squatter settlements. This adds to the vulnerability due to disasters.

Gujarat is one of the most economically advanced states in India. As of 2000, Gujarat was ranked 4th nationally in per capita income (Indian Rupees, Rs. 13,298)

Table 6.3 Key demographic details of India, Gujarat, and Kutch

	India		Gujarat		Kutch	
	2001	2011	2001	2011	2001	2011
Population	1.028 billion	1.210 billion	50.671 million	60.383 million	1.500 million	2.09 million
Population density	312.9	382	258.4	308	35	46
Urban population (%)	27.8	31.16	37.4	42.6	30.0	34.72
Literacy rate (%)	65.38	74.04 (82 men/ 65 women)	69.97	79.3 (87 men/ 70 women)	59.79	71.58 (80 men/ 61 women)

Source: Census of India (2011)

Table 6.4 Town profiles (2008–2009)

Towns	Total population	Class	Area (sq km)	Slum population (# of slums)	% slum population to the total population	Municipal staff
Bhuj	123,536	B	36	52,722 (38)	43	512
Mandvi	52,984	C	15	8,700 (13)	16	193
Bhachau	36,444	D	9	11,800 (7)	32	84

Source: CEPT University (2012)

and 11th in percentage of people below poverty line (14 %). Gujarat has witnessed fast growth since 2000 and now ranks second in per capita income (Rs. 63,549) nationally (Census of India 2011). Compared to other Indian states, Gujarat was able to invest in disaster risk management after the 2001 earthquake because of the availability of funds and resources (Erramilli 2009). According to Erramilli (2009), Gujarat had the most elaborate administrative mechanisms for dealing with disasters, compared to other states such as Andhra Pradesh, Orissa, and Bihar, due to its comprehensiveness and access to resources. His study found that Gujarat state was able to undertake a large procurement exercise of:

> equipment that ranged from simple to advanced and heavy machinery, which were useful in emergencies. These ranged from life jackets, ropes, inflatable tubes, polythene sheets to heavy earth moving equipment, sophisticated steel-cutting and concrete-cutting machinery. (p. 115)

Kutch has an important role in the economic development of the state as well as the country. Kandla port, the largest seaport in India, is in Gandhidham, on the Gulf of Kutch. Mundra is developing into another important port on the west coast of India. Major agricultural products include oil seeds, food grains (bajra and jowar), cotton, pulses (legumes), and date palms. There are large deposits of minerals that include gypsum, lignite, lime stone, and bauxite. Salt production is another important economic activity. There are a handful of manufacturing and processing industries as well as numerous small-scale industries in the district, many of which were established after the earthquake as part of the state policy that gave tax breaks for industries setting up operations in Kutch. Tourism is an important economic generator attracting people to the palaces and archeological and pilgrimage sites. Kutch is also famous for its handicrafts including ornaments and clothes. The town of Bhuj is the administrative headquarters of the district. Its economic base comprises government, agro-based business, services, handicrafts, and tourism. Bhuj also has a large military presence as the district is close to the Pakistan border. In the aftermath of the 2001 earthquake, the town has seen an increase of service-based industries. Bhachau was traditionally a small town with an agro-based economy. However, the new industrial policy of the state government resulted in many large industrial plants being established near the town. Mandvi has traditionally been a center of tourism-, port-, and agro-based activities and remained the same even after the earthquake.

6.4.2 Public Administration, Democratic Decentralization, and Politics

In order to understand the capacity of government to deal with disasters, it is important to understand the prevailing governance and administrative structure. The administrative system in India is very hierarchical though the local governance system has some autonomy under recent constitutional amendments. The strengths and weaknesses of this system contribute to the region's vulnerability to disasters. Though the system has reasonably good reach to all levels of administration such as district, taluka, and village, the functioning is questionable. This could be attributed to the structure, decentralization of functions, and the general political climate, which also affects the capacity of government agencies. The interviews and surveys attempted to understand the capacity of government and the community. In order to understand the interview and survey responses, it is first necessary to understand the administrative structure and politics.

India, a federal union of states, comprises 28 states and 7 union territories. The President is the constitutional Executive of the Union. Real executive power vests in a Council of Ministers with the Prime Minister as head. The Council of Ministers is collectively responsible to the Parliament, which has two houses (Rajya Sabha, the upper house, and Lok Sabha, the lower house). Within the states, the Governor, as the representative of the President, is the Executive, but real executive power rests with the Chief Minister who heads the Council of Ministers. The Council of Ministers of a state is collectively responsible to the elected legislative assembly of the state. Union territories are ruled directly by the federal government through a Lieutenant Governor, appointed by the President. The Constitution governs the sharing of legislative power between Parliament and the State Legislatures and provides for the vesting of residual powers in Parliament. Part XI of the Indian constitution defines the power distribution between the federal government (the Center) and the states in India. This part is divided between legislative and administrative powers. The legislative section is divided into three lists: Union list (authority of the center), States list (authority of the state), and Concurrent list (authority shared by the center and states). Disaster management comes under the Concurrent list entry 23, social security and social insurance (Government of India 2007).

Gujarat State is divided into 24 districts, and each district is further subdivided into *talukas*. Each of these *talukas* has a number of villages. For administrative purposes, Kutch district has been divided into nine *talukas*: Bhuj, Anjar, Nakhatrana, Lakhpat, Mandvi, Mundra, Bhachau, Rapar, and Abdasa. As per the 2011 census, there are a total of 933 villages and 6 towns in the district (Census of India 2011).

The public administration all over India is dictated by a hierarchical bureaucratic system and a parallel elected representative system. The bureaucratic system is a modified version of the British system that prevailed before independence (in 1947). Many of the institutions and officers have distinctive Indian names (Taluka, Mamlatdar, etc.) used in both English and Hindi discourses on Indian local government. The local government system, which existed from independence, received more autonomy in 1992 as per the 73rd and 74th constitutional amendments. The bureaucratic system is headed at the state level by the Chief Secretary who is supported by a number of Additional Chief Secretaries and Principal Secretaries responsible for various government departments such as urban development, rural development, education, health, and law. These high-level officials are from an elite national Indian Administrative Service cadre (called IAS officers). These departments have state level, district level, and taluka (subdistrict) level offices to manage the activities. At the state level, the state legislature, headed by the Chief Minister and the cabinet of ministers, lead the administration. The bureaucratic officers are under the cabinet.

One of the most influential departments is the Revenue Department. Its head at the district level is the District Collector (called Collector here onwards). The Collector is also an IAS officer. The Collector has three additional responsibilities. The first is that of District Magistrate (DM), looking after law and order. Though the police are under the Home Department, the Collector has supervisory authority over them at the district level. The second is that of Crisis Administrator, managing and coordinating activities during natural and man-made disasters as well as social unrest. The third role is that of Development Officer, heading the rural and economic development of the district. Thus, disaster management is one of the Collector's primary responsibilities.

For many state government bodies, the Collector had an important role in determining how, where, and what quantity of their services were to be delivered. This structure has seen changes from the late 1990s, after the introduction of the Panchayati Raj system (local empowerment) in the country in 1993. Until the 1960s, the Collector was the all-encompassing powerful institution taking care of all governmental programs. With the introduction of a large number of new projects/schemes in the 1970s, the Collector's role was limited to periodic review/monitoring of various departmental programs. Initiation of a large number of developmental activities and programs throughout the country in the 1980s saw government departments and agencies carrying out the day-to-day activities. Still, the Collector remained the omniscient and omnipotent source for solutions to any potential troubles. However, after the introduction of the Panchayati Raj Institutions (PRI) in the country in 1993, most of the development functions have been taken away from the Collector's domain, although the State Governments use this institution to extend their reach.

The Panchayati Raj, or local self-government in India, is a three-tier structure below the states, with Gram Panchayat (village local body) at the lowest level, Taluka Panchayat (Block Panchayat) at the intermediate, and Zila Panchayat (District Panchayat) at the district levels (Sheth 2000). Compared to other states in India, Gujarat has done relatively well in developing Panchayati Raj Institutions. Gujarat state has also provided a lot more resources to Panchayati Raj Institutions compared to other states (Sheth 2000). In the pre-1992 period, it was reported that the state had transferred about 25 % of its revenues to various panchayat institutions and in 1977–1978, their per capita income was Rs. 9 (nine Indian Rupees). The comparable amounts in other states were, for example, Rs. 0.78 in Uttar Pradesh and Rs. 1.25 in Orissa. However, Panchayati Raj Institutions in Gujarat were not interested in raising their own revenues through taxes and fees and were reluctant to exercise their powers. Therefore, it was estimated that about 98 % of funds were transferred by higher tiers, and PRIs raised only 2 % from their own sources (Erramilli 2009). Thus, although PRIs were relatively successful in Gujarat, they did not function as autonomous bodies of governance.

In Gujarat, the current administrative structure is as presented below. The Collector is now supported at the district level by Additional Collector, Deputy Collectors, and Revenue Officers. The Collectorate coordinates activities of the line departments at district level. The District Development Officer (DDO) acts as the link between the Collector and the district level elected representatives at the District Panchayat. There are 33 elected representatives at the Kutch District Panchayat with one of them being selected by the majority party as the President. The Taluka (Block) is the intermediate level of administration between the district and village. Taluka administration is headed by the *Mamlatdar* under one of the Deputy Collectors in charge of the Taluka. The Mamlatdar is supported by Deputy Mamlatdars. The Taluka Development Officer (TDO) acts as the link between the Taluka administration and the Taluka Panchayat (Block Panchayat), the elected body at Taluka level. There are 9 Talukas with an average of 18 elected representatives at each Taluka Panchayat in Kutch (Bhuj, being the largest taluka, has 27 representatives). Similar to the District Panchayat, one of the elected representatives from the majority party becomes the President of the Taluka Panchayat. At village level, the Talati is the administrative head, doubling up as the link to the elected body of the village panchayat. There are a total of 615 village panchayats in the district, with each having an elected body headed by a *Sarpanch*. Figure 6.6 below presents this hierarchical structure in graphic form.

The urban area administration is under separate entities called municipal corporations or municipalities (Classes A to D) based on population. While Bhuj is a Class B municipality, Mandvi is a Class C municipality, and Bhachau is a Class D municipality. Though municipalities are autonomous bodies, administratively they

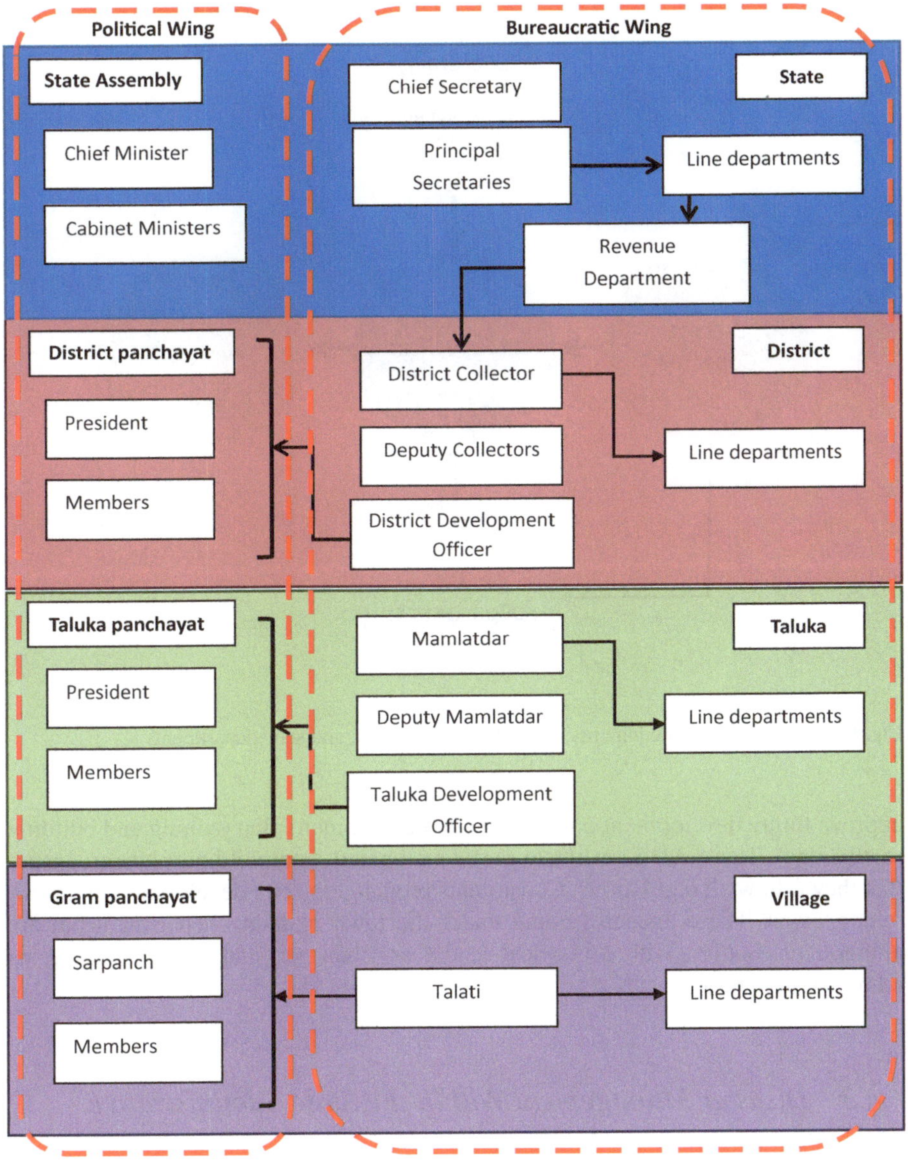

Fig. 6.6 Administrative system in India

come under the state Urban Development Department. The Chief Officer is the bureaucratic head of the municipality, linking it with the state and district administrations. A body of elected representatives is headed by the President (Mayor) who is selected by the majority political party (Fig. 6.7).

After the earthquake in 2001, metropolitan area development authorities (ADAs) were established in the major towns of Kutch to plan and reorganize the towns and

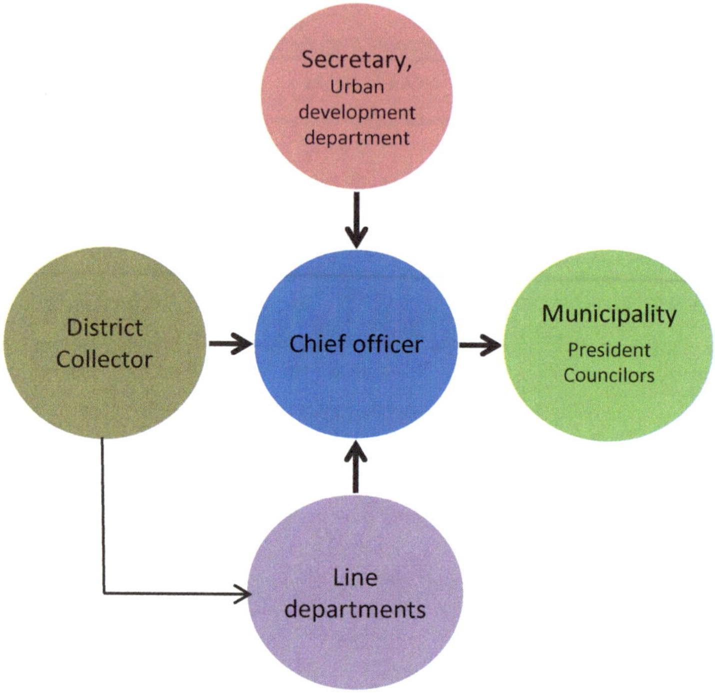

Fig. 6.7 Link between municipality, district administration, and state departments

approve future development based on a city development plan (zoning and building regulations). These ADAs evaluate plot development and building plans to ensure that they follow all regulations. Consultants prepared the city development plans for major towns. These agencies come under the town planning department but are autonomous bodies, with a political leader as Chairman and a town planner as administrative head.

6.4.3 Disaster Management Within the State Administrative Structure

Comprehensive reform of disaster management field was taken up by the federal government after the formation of GSDMA in Gujarat in 2001. The Ministry of Home Affairs was notified as the nodal (central) ministry for Disaster Management in 2002 (Government of India 2004). An Administrative Reforms Commission examined and suggested measures for efficient and sustainable administration in all fields including disaster management (Government of India 2005). A comprehensive Disaster Management Act was passed by the Indian Parliament in December 2005; this act established the National Disaster Management Authority (NDMA).

The NDMA spearheads and implements a holistic and integrated approach to disaster management. The act encouraged states to set up Disaster Management departments and Disaster Management Authorities to promulgate Disaster Management Codes (The Gazette of India 2005). Since the Gujarat State Disaster Management Authority (GSDMA) was established in 2001 and GEERP was initiated soon after (before national agencies were set up), GSDMA had the sole responsibility and control over the GEERP activities. The current role of national agencies is to support state agencies and coordinate interstate activities.

The reforms also made the Indian Meteorological Department, the Central Waters Commission, National Core Groups on Landslide Mitigation and Earthquake Mitigation, and the Bureau of Indian Standards responsible for developing and updating early warning systems and zoning and building regulations (NDMA 2007). The National Center for Disaster Management, which was set up in 1995, was upgraded to a National Institute of Disaster Management (NIDM) in 2003 with the purpose to conduct research, undertake documentation, develop training modules, conduct training programs, and assist training institutions and state institutes. After promulgation of the Disaster Management Act in 2005, NIDM was recognized as a statutory, nodal institution. NIDM coordinates its activities with the state counterparts.

Prior to the earthquake in 2001, disaster prevention was not considered seriously in Gujarat. The relief and reconstruction was coordinated by the Collector at the district level or by Deputy Collector and Taluka Development Officer (TDO) at the Taluka level. The municipalities or other local bodies carried out immediate responses. No municipalities or local bodies were well prepared for quick response due to lack of equipment, as well as a lack of clear roles and responsibilities. It was left to the administrative and political leaders to coordinate and manage the activities. The majority of the responses were for fires and local flooding. Major disasters such as cyclones that affect the district regularly were left to the district administration to deal with. Since cyclones can be predicted reasonably well in advance, national and state meteorological departments used to give warnings. The district administration would take actions such as giving warnings to the fisherman and others living in coastal zones and evacuating the most vulnerable in extreme conditions. This was particularly true during the 1998 and 1999 cyclones. NGOs were also involved in response activities. However, they were more focused on drought, a persistent problem faced by the entire district for a long time.

The intensity of the 2001 earthquake was extensive both geographically and in terms of damages. Many local bodies were left helpless as entire towns and villages were affected, with many emergency responders themselves affected and not able to attend to their duties. Many government buildings including the Collectorate were damaged severely. The state government sent special officers (IAS officers on special duty) to lead the rescue and relief work. They, along with the Collector, coordinated the activities of state/local governments, international/national aid agencies/NGOs, and community/religious organizations. NGOs coordinated volunteers from different parts of the state/country for relief work.

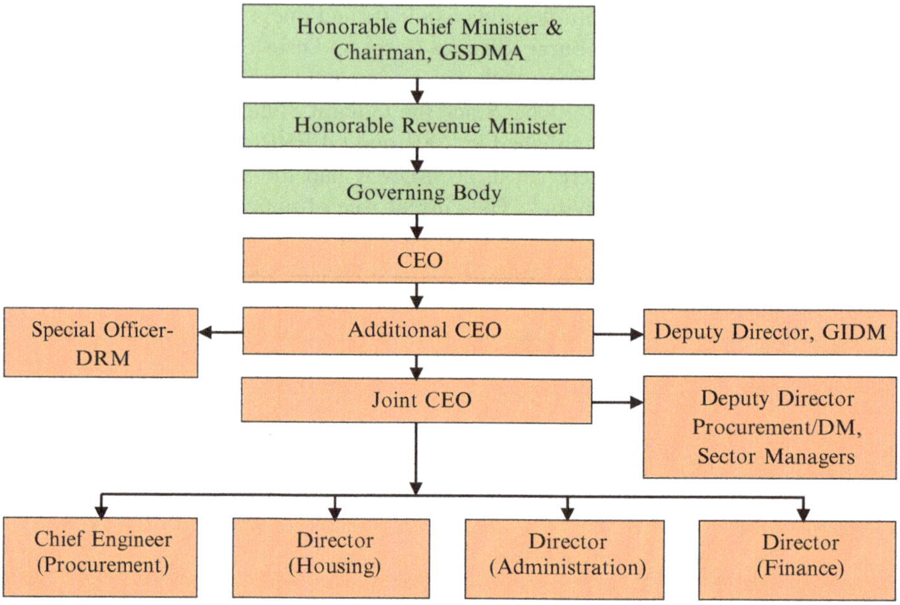

Fig. 6.8 Administrative structure of GSDMA (Source: GSDMA (2012))

A State Disaster Risk Management policy was established in 2001, right after the earthquake, and was followed by the enactment of the Gujarat State Disaster Management Act. This act established the GSDMA as the nodal agency for disaster relief and rehabilitation, preparedness, and risk reduction. Thus, Gujarat established state laws and a nodal agency before the central government established laws and agency at the federal level. The authority is headed by the Chief Executive Officer (CEO), appointed by the state government from the national IAS cadre. The CEO is supported by an Additional CEO, a Joint CEO (disaster management), a Chief Engineer (procurement), and three directors (housing, administration, and finance). Figure 6.8 above shows the organizational structure of GSDMA at the state level. This is further extended to the district level through the District Emergency Operations Center (DEOC). Initially directly under the District Collector, the DEOC now works more independently, though the Collector still has the supervising power and assumes full control during an emergency. DEOCs have appointed qualified personnel as staff over the years and coordinate GSDMA activities at district level. They also supervise activities at the taluka and village levels through TDO and Talati.

6.4.4 Politics

Before coming under the British rule in the early nineteenth century, Kutch was under a succession of kings. After the independence of India in 1947, Kutch became part of Bombay state. It became part of the Gujarat state in 1960 when Bombay state was divided into Maharashtra and Gujarat. The Indian National Congress (INC), which played a key role in Indian independence, won elections in the state until 1995, relying on its vote banks and coalitions (Yagnik and Sud 2005). Thereafter, the right wing Bharatiya Janata Party (BJP) won every election in the state in 1995, 1998, 2002, and 2007 (Erramilli 2009). BJP retained its rule in the state for another five years by winning the election in December 2012. Both political parties, INC and BJP, have widespread support in the Kutch district. While the INC has high influence in the rural areas, the BJP grabs influence in the urban areas. In the 1998 elections for the Gujarat Legislative Assembly, the BJP had 45 % of the votes, and the Congress 35 % of the votes. In elections after the 2001 earthquake, the BJP again won the elections, gaining even more votes (50 %). The two leading political parties did not use disaster management as an important issue in the elections (Erramilli 2009).

6.4.5 Community Structure

Similar to the link between governance structure and government capacity, the capacity of a community to deal with disasters is linked to its sociocultural structure. The strict hierarchy of Indian society has a large influence on social dominance, residential location choices, economic activities, and political and administrative influence. During the field work, it was evident that this community structure has important implications for communities' capacity to deal with disasters.

Kutch is inhabited by various groups, who have migrated over centuries from regions of western Rajasthan, Sindh, and even Afghanistan. These include nomadic and seminomadic groups who are mainly cattle herders. The majority of the population follows Hindu religion, but a considerable number of people follow Islam and Jainism as well. Within the Hindu community, there are many caste groups (social hierarchy), just as in other parts of the country. The highest level groups include brahmins, darbars (rajputs), and jains (luhaanas). Major peasant groups include patels, ahirs, and rabaris. The bottom level (in social hierarchy) includes artisan groups, scheduled castes, and scheduled tribes. Most of Kutch still follows a feudal social order with the feudal elite dominating social, political, and economic life (Katiyar and Khandelwal 2001).

Many religious and caste groups in the higher echelons have their own religious or caste-based organizations that support them in time of need. This was evident in the aftermath of the 2001 earthquake when economically powerful and well-connected groups (connected to similar organizations outside the district) could generate material and financial support in a short time while others were left on their own. Some of the major community-based organizations (CBO) include the Swaminarayan Trust and the Jain Samaj. In addition to such CBOs, Kutch has seen committed and long-term activities by a number of nongovernmental organizations (NGO). These NGOs either work on a number of development disciplines in a focused geographical area or on a few specific disciplines throughout the district.

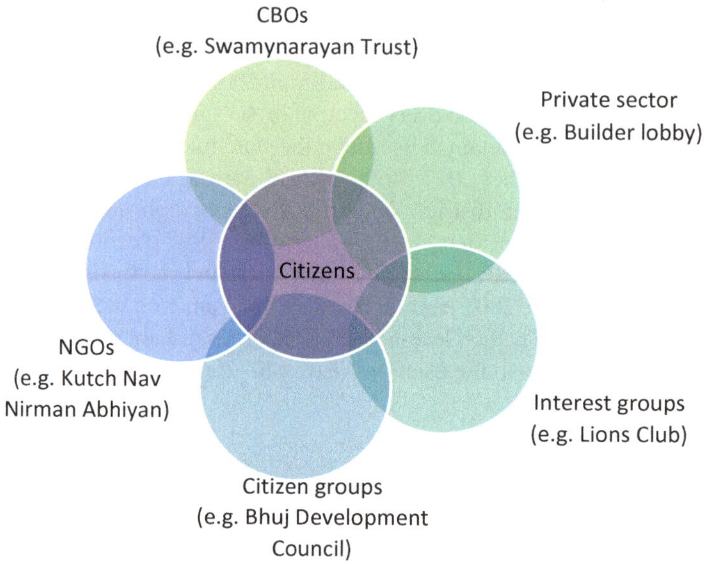

Fig. 6.9 Community structure in the study towns

After the cyclone of 1999, a number of NGOs came together to establish Kutch Nav Nirman Abhiyan (KNNA), a collective of Kutch-based development organizations with a strong local presence. Currently, KNNA has more than 30 members working on social empowerment, advocacy, and facilitation of collaboration among communities, government, donors, and NGOs. There are also other interest groups such as the Lions Club, Rotary Club, and (building) Developers' Association. These groups also take a keen interest in community support activities. Citizen groups such as the Bhuj Development Council (BDC) were started as social get-together groups of retired engineers and professionals who took up the active role of mediating between community and government officials after the 2001 earthquake. They also coordinated community meetings with citizens, government officials, and consultants during the redevelopment. In this research, I consider community in its general sense where all the abovementioned groups are included (see Fig. 6.9).

6.4.6 Built Environment and Urban Development

The built environment has a large influence on disaster vulnerability. Settlements planned with disasters in mind could reduce the impacts, while haphazard development could lead to heavy damages to properties and deaths. Prior to the 2001 earthquake, existing town planning rules and building regulations had been ineffectively enforced and largely ignored by the residents in the study towns. Many buildings were illegally constructed beyond the allowed coverage (percentage of the plot area which can be built) and height (number of floors). A large number of deaths and injuries occurred in the densest areas and high-rise apartments in Kutch. In Bhuj, for example, large number of deaths and severe injuries occurred in the walled city (high density with narrow streets) and high-rise apartments (poor quality construction and/or violation of regulations).

In response to the 2001 earthquake, the government announced a policy of limiting new building construction to only two floors for a 5-year period. They also announced a range of other measures that included limiting building heights to seven meters and enforcing a tighter Floor Space Index (total floor area to land cover ratio) limits. The new development regulations allocated significant land in the urban centers to infrastructure and open space, partly with the objective of decreasing population density and increasing access of emergency vehicles during disasters. All these measures had the overriding objective of improving the safety in urban areas. However, they also contributed to a substantial increase in the extent of urban sprawl in towns such as Bhuj.

The GEERP included land-use planning for affected urban areas. Area development authorities (ADAs) were set up in these urban areas to oversee land-use plan development and implementation. The ADAs were also in charge of a "build back better" strategy, which focused on building disaster resilient housing and infrastructure. The ADAs were successful in decongesting urban areas and improving infrastructure quality. For example, the area under the Bhuj Area Development Authority (BHADA) has almost doubled in the 12 years following the earthquake (2,351–4,239 Ha). The majority of this increase is in residential use with substantial increases in industrial, commercial, public, and transportation uses. The urban spatial structure of the Bhuj town has changed drastically from the pre-disaster to the post-disaster periods. While residential uses almost doubled, commercial, industrial, public, and recreation uses also saw minimal increase. More roads have been added to the system. Some water bodies have been restored. As a result of this urbanization, some agricultural uses have been pushed out of the metropolitan area. Besides the changes in the land-use pattern, the improved road network and better linkage of Bhuj with surrounding towns have contributed to the urban growth of the town in the southern and northern directions, converting a large share of agriculture and reserved lands in the north to residential and industrial lands. The development of a hierarchical network of roads with well-planned streets, trunk roads connecting the walled core area and outer areas, and ring roads looping the urban areas under the control of the planning authority significantly increased the level of amenities and land quality of the town (BHADA 2011). See Fig. 6.10 for changes in the road network and plots as a result of the reconstruction planning.

The government has provided a number of incentives for decongesting towns and developing a better urban landscape, such as: allotment of developed lands larger than the lands lost through land adjustments (Town Planning Schemes); financial and technical supports for constructing new houses or repairing damaged houses; supplementary assistance by donors and NGOs; and concessional housing

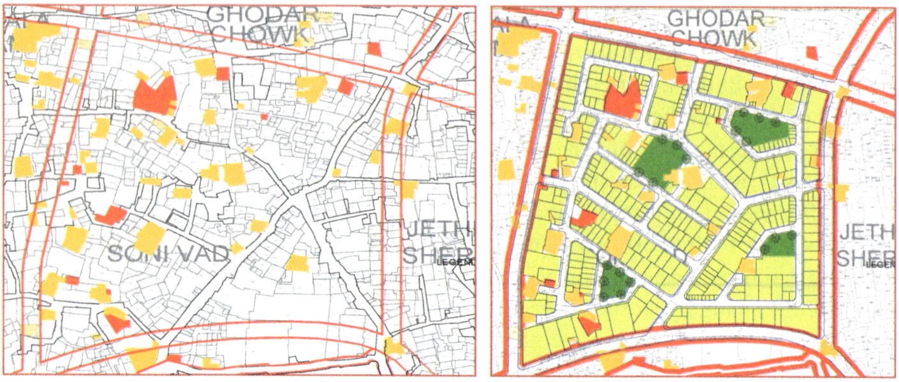

Fig. 6.10 Road network and plots in the central area of Bhuj in 2000 (left) and 2009 (right) (Source: BHADA (2011))

loans and housing insurance. Houses and infrastructure, in principle, are more earthquake resilient than before. However, scholars have pointed to the continued vulnerability of houses constructed under owner-driven reconstruction programs due to less technical expertise in the construction stage (Jigyasu 2002a). The program also improved the accessibility of families resettled in urban fringes by relocating some of the public offices and expanding the urban infrastructure and services to the newly developed areas. In addition, the movement of tenants from the town centers to the urban fringes and the revised building bylaws also contributed to urban sprawl by pushing families to the newly developed suburbs. Since most of these families were given the option to choose the relocation sites and were grouped into community-based clusters, these developments have also contributed to increased spatial segregation of communities based on castes and religions (Simpson 2008).

The study area faces severe risks from multiple natural hazards such as earthquakes, cyclones, floods, and droughts. The capacity of government agencies and communities to prepare for and respond to natural disasters has been influenced by its socioeconomic vulnerability factors. The environment and disaster risk in the study area, presented above, provides an important backdrop against which the findings from interviews and surveys are discussed below.

6.5 Findings

6.5.1 Risk and Capacity Perception

How do different actors perceive the capacity deficit just after the earthquake and now? This section addresses the topic of perception, looking at how citizens in the three towns understand the capacity deficit and respond based on those understandings. It draws from the interviews and surveys carried out during the field work in 2012. The assumption underlying this question is that different groups have different perceptions of current capacity, capacity deficit, and disaster risk. Their capacity development efforts are linked to their perception of capacity deficit and disaster risk.

As per the conceptual model, governments, donors, and communities take actions based on their perception of current capacity and capacity deficit as well as their understanding of disaster risk (see Fig. 5.4, which presents the PeDJoLA conceptual framework). The field study, donors' and government's perceptions are understood through various donor- and government-led assessments, which formed the basis of the GEERP, as well as through interviews and surveys. Community members' perceptions were gathered through formal and informal interviews and surveys.

The activities carried out after the 2001 earthquake, particularly the implementation of the GEERP activities, were expected to increase the capacity of the government as well as the community groups and residents to understand and reduce the risks, increase the disaster preparedness, and increase the capacity to deal with future disasters. The before-and-after analysis is an important indicator of the changes that happened in the government structure, the reach of the programs, and the internal changes in the community. Both interviews and structured questionnaire surveys started with questions about the 2001 earthquake and perceptions regarding the situation before and after.

6.5.1.1 Interview Responses

Interviews provided multiple viewpoints as the interviewees included government officers, retired government officers, NGOs, interest groups, CBO representatives, and city development experts. Most of these respondents had been local residents and were involved in the rehabilitation process in Kutch after the 2001 earthquake. They know the community dynamics and the intensity of government and community activities. While talking about their experiences and knowledge of the situation during and immediately after the earthquake, they reported the risk perception and capacity before the earthquake and how these changed over time.

Different groups perceive capacity differently. However, there is a clear division of opinion between government and community respondents. Among government officers, perceptions of state and district government varied. Government officers at the state and district levels firmly believe that the capacity of government institutions has increased considerably. They attribute this to the establishment of GSDMA and its activities. The state of Gujarat enacted the Disaster Management Act and set up a State Emergency Operations Center (SEOC) and District Emergency Operations Centers (DEOC). As one of the GSDMA officials mentioned:

> We have received numerous awards for GSDMA; we have set up a system that extends all the way down to the village level; we have established various programs and technical com-

mittees that recommend new programs or modification to existing programs, training, technical studies, and mapping.

It seemed that the state and district level officials were concerned more about capacity of government than of the community itself. With respect to disaster preparedness, they believe that the protocol for early warning, training provided to government staff, and awareness programs for citizens make both the government and the community better prepared for facing a future disaster. It is also their belief that the post-earthquake reconstruction and the activities of area development authorities have significantly reduced the risk of disaster damages.

Respondents from municipalities believe that their capacity has increased but that their role focuses only on rescue and relief. For example, a municipal staff commented:

> Our responsibility is to respond and provide relief. We received training for that and we definitely increased capacity on that aspect.

They also believe that they are better prepared to face disaster now than before. They attribute this to the past experience as well as training and the new institutional setup that guides disaster management. However, they also point to the fact that the new development is regulated by the area development authorities (ADAs) and they do not have a direct role in regulation.

> Municipality does not deal with building permits and construction these days. It is done by ADA. We are responsible for relief work. We will do as much as we can. (Municipality respondent)

ADAs are confident that they are doing a good job of city planning and building regulation. However, they point to illegal construction that might still pose risks in future disasters and note that they do not have the power or resources to carry out enforcement.

Municipalities believe that citizens are better aware of the disasters through experience and awareness programs carried out under GSDMA. However, as one municipal respondent put it:

> The memory of people is short and we need to carry out programs more frequently.

Respondents from outside the government included representatives from NGOs, CBOs, interest groups, and political leaders (see Appendix). They have a different perspective and mostly mentioned that there was no real capacity for disaster management before the earthquake as was evident from the damage caused by the earthquake. They feel detached from the government activities and do not really know how much capacity has changed within government after the earthquake.

> We don't know what the government is doing. (NGO)
> Authorities [new agencies] are set up but they do not involve us in the process and so we don't know. (NGO)
> I have heard about the State Act and agencies but they are not well publicized. (Retired government officer)

They believe that the capacity and preparedness of citizens are still low. This is attributed to the lack of awareness and the relaxed attitude. In communities that are

Table 6.5 Perceptions of different groups based on interviews

Group	Disaster risk 2001/2012	Capacity 2001/2012	Capacity deficit 2001/2012
Donors/state/district	Low/high	Low/high	High/low
Municipality and area development authorities	Low/high	None/low	High/high
NGOs, CBOs, technical and academic experts	Low/high	None/low	High/high

better prepared, preparedness is attributed to the past experience of people rather than any government action.

> There is a severe lack of awareness among citizens about the do's and don'ts before and during a disaster. (NGO)
> If they are better prepared, it is not because of government but because of their bad experience in the past. (Municipal Councilor)

Respondents from outside the government all consider that the risk is still high as it was before the earthquake. The only consolation they think is that the deaths will be less in the future as the towns are now decongested and have wide roads and open spaces.

In summary, the perceptions of groups differ. The state and district administration argue that the new institutional setup and programs have significantly increased the capacity of government, improved preparedness, and reduced risk. The municipalities and ADAs are responsible for many aspects of response and preparedness. They are not entirely sure if their capacity is adequate or if they are well prepared. Community groups and citizens believe that the government capacity to respond has increased but not necessarily capacity on preparedness. They point to the lack of direct links between government actions and citizens as well as community organizations (Table 6.5).

6.5.1.2 Survey Responses

The survey included government staff involved in disaster management as well as a small sample of residents in Bhuj (see Appendix). About 90 % of the respondents from government agencies across the three towns agree that the government capacity to deal with disasters was low before the earthquake. Out of all the respondents, about 93 % also believe that the government is better prepared now than before. Compared to other study towns, respondents in Bhuj have the lowest agreement (77 %) on this assertion. This shows that the perception of preparedness has improved across the study towns. While there is no difference in opinion among different agencies, most of the ADAs restrained themselves from offering any opinion on the capacity deficit before the earthquake as ADAs were formed after the earthquake. Table 6.6 below shows the overall percentages.

While all Bhuj residents surveyed agree that the capacity to deal with disasters was low in 2001, there is less agreement in perception about preparedness. Only 68 % of them agree that they are now better prepared than before. About half of

Table 6.6 Disaster risk perception of government staff

		Percent	Number of responses
Capacity in 2001 was low	Agree	82	26
	Disagree	6	2
Better prepared now	Agree	94	30
	Disagree	6	2

Table 6.7 Perception of residents of Bhuj

		Percent	Number of responses
Capacity in 2001 was low	Agree	100	50
Better prepared now	Agree	68	34
	Disagree	32	16
Know safe building techniques	Agree	54	27
	Disagree	46	23

those surveyed (54 %) agreed that they are aware of safe building techniques. However, the rest had no idea what the safe techniques are. Knowledge of safe building techniques was used as a proxy for preparedness. Even if we assume that awareness of safe building techniques would have translated into better buildings over the years, the surveys show that only about half of the surveyed residents in Bhuj are prepared for future earthquakes. While the residents in squatter settlements know safe building techniques, the majority of them considered themselves no better prepared than before. This could be attributed to the low-income levels in squatter settlements that prevent the residents from making costly housing renovations. This is evident from the fact that, generally, respondents from low-income group reported that they are not better prepared now than before. An analysis based on gender did not show any differences. While both males and females agree that the capacity was less in 2001, they think that they are better prepared now. There was slightly less agreement among females about the knowledge of safe building techniques. This could be attributed to the prevailing social system that gives fewer roles for women in building construction and related decision-making. Table 6.7 above shows the overall percentages.

6.5.1.3 Perception of Capacity and Risk Differ Now

Overall, the field research shows that different groups perceive disaster risk and capacity in different ways, during different times. Government staff at state, regional, and local levels as well as surveyed residents believed that the capacity of both government and community as well as the perception of earthquake risk was low before the 2001 earthquake. Both government and community perceive that they are better prepared in 2012 to deal with disasters. However, they agree to different degrees:

- Donors, state/district governments—low capacity deficit/low risk
- Local government (municipality and ADA)—medium capacity deficit/high risk
- NGOs, CBOs, technical and academic experts—high capacity deficit/high risk

Bhuj residents that were surveyed agreed that the capacity to deal with disasters was low in 2001. Within the residents surveyed in the city of Bhuj, there are minor differences of capacity perception depending on the economic status and location of residents. Generally, low-income residents and those residing in squatter settlements perceive less capacity development and preparedness than others.

The next section links capacity perceptions with capacity development interventions.

6.5.2 Forms of Capacity Development

What were the different forms of capacity development undertaken in Gujarat after the 2001 earthquake? This question relates to the design of interventions in the conceptual model and is divided into three sub-questions: (i) What are the different forms of capacity development undertaken in Gujarat after the 2001 earthquake? (ii) Were the adopted capacity development forms based on perceived capacity deficits (as discussed in the previous section)? and (iii) What mechanisms were used by donors and the government to formulate intervention designs, implement actions, and monitor the results? The basic premise of this part of the conceptual model is that different groups undertake capacity development efforts based on their perception of capacity deficit and disaster risk. This section relies primarily on the analysis of secondary data from donor assessments, government reports, media, and past studies and secondarily on analysis of interview data.

6.5.2.1 Forms of Capacity Development

Three broad forms of capacity development took place under the GEERP. The first form was to establish new organizations and develop a new institution. The second form applies to the idea of capacity development as reconstruction. This form is very common in a post-disaster situation and is based on the concept of "build back better." The third form of capacity development under GEERP was training, skill building, educational courses, and awareness programs. Overall, the project was conceived and implemented in a very top-down manner.

New Institutions

Within weeks after the 2001 earthquake, the state government in Gujarat set up a special body, the Gujarat State Disaster Management Authority (GSDMA), to deal with all aspects of relief and rehabilitation. Headed by an IAS officer, GSDMA worked with special officers at the district headquarters to coordinate relief activities in the early stages. This agency was conceived to be a permanent arrangement to handle natural disasters (see Sect. 5.1, subsection titled "Disaster management within the state administrative structure" for details about GSDMA). The Gujarat Institute of Disaster Management (GIDM) was established as a training and research wing of GSDMA on January 26, 2004, by the government of Gujarat with the aim of human resource development in the state. Its objectives include providing disaster management training, undertaking public education and community awareness, acting as a resource center and clearing house of information, and facilitating partnerships with private organizations and universities. Currently, GIDM offers a series of training courses to government officials and other stakeholders. These courses are offered by experts in the field. Four area development authorities were established in the four affected towns of Kutch to develop, coordinate, and implement urban development plans. These area development authorities are now responsible for providing building code permission and enforcement. Many of these functions were with municipalities before 2001.

Reconstruction: Development Planning, Relocation, Rehabilitation, and Repair

The GEERP used a "build back better" reconstruction approach to ensure that the new infrastructure and buildings allow for emergency response and earthquake-resistant construction. "Build back better" is a principle promoted by the United Nations and other international development agencies, suggesting that the reconstruction should be geared toward longer-term sustainable development in terms of better housing, education, health care, and economic opportunities (UNISDR 2011). In Kutch, this involved large-scale consultant-driven urban development planning and implementation through the newly created area development authorities. New urban development plans provided the basis for reducing density and planning for large-scale resilient infrastructure and building construction. Traditionally, municipalities were responsible for development planning and enforcement. However, as they had limited technical and financial capacity, the reconstruction process was coordinated by the newly established area development authorities with the help of consultants.

A number of NGOs were involved in reconstructing houses. A total of 74 NGOs, including 10 international, 20 national, and 26 local NGOs, participated in the reconstruction of 42,000 houses (Thiruppugazh 2007). NGOs had a significant input into the housing rehabilitation and reconstruction program. Of the 209,915 houses that had been reconstructed to the end of March 2006, 41,902 (nearly 20 %) had been constructed by NGOs and the private sector. NGOs provided significant support to houses assessed in the worst categories (G-1 and G-2) with respect to damages (UN-HABITAT 2009). However, some researchers have mentioned that many NGOs did not involve the local people in decision-making (Jigyasu 2002a) and some withdrew after physical reconstruction without looking into the long-term sustainability of the interventions.

Media and researchers have also noted that government's earlier recovery and reconstruction plan was based on relocation of people and housing, which was met with stiff resistance from the local people who did not want to be uprooted. As a result, the government decided not to press for relocation and advocated for "owner-driven" reconstruction as its primary approach (Jigyasu 2002b). The government agreed to provide financial assistance to all those who did not want to relocate. Such beneficiaries were supposed to undertake reconstruction on their own.

Training, Education, and Awareness

Capacity development efforts under this form were primarily aimed at increasing capacity by training masons and semiskilled laborers in earthquake-resistant construction techniques, introducing education courses on seismic engineering in universities, and developing awareness programs in schools and through media resources. Training government staff in relief and emergency response procedures also emerged as a primary capacity development effort. However, many interview respondents underscored that the training and skill building programs were not primarily based on the need and local context (demand) but more on decisions made at GSDMA headquarters or by GIDM (supply).

The GSDMA has established a number of training programs for government staff and interested NGOs and private groups. The topics of the training programs vary from emergency response (e.g., firefighting) to safe building technologies and are intended to develop and improve various skills. GIDM has been set up under the GSDMA with a pool of full-time and adjunct faculty to conduct training programs. Most of the training is currently carried out at the Sardar Patel Institute of Public Administration in Ahmedabad, the largest city in the state. A separate campus for GIDM is being developed near the state capital Gandhinagar. The programs usually run for five days on alternate weeks, the schedule of which is sent to concerned agencies in advance for registration. By the end of 2012, about 250 programs have been conducted in nearly 10 years and have trained about 6,300 people across the state. These include staff of municipalities, line departments, area development authorities, private enterprises, and NGOs.

Awareness programs are another important aspect of capacity development where the government deals with the citizens and NGOs/CBOs. The DEOC has responsibility for conducting awareness programs at district, taluka, and village levels. They carry out a number of activities such as displays of posters at prominent locations in agencies where people can easily see them, distribution of pamphlets, response/safety demonstrations (e.g., by the fire department), movies, competitions for school/college kids, drills, and rallies. Seasonal advertisements are aired on television channels and local cable networks during prime time. Special occasions such as "Fire Safety Week" and "UN Disaster Reduction Day" are celebrated (Table 6.8).

Table 6.8 Training, education, and awareness programs under the GEERP

Focus areas	Actions	Comments by interview respondents
Seismic engineering design skills	Syllabus changes in civil engineering and architecture programs, teacher training programs, continuing education programs for practicing engineers in private and government sector	Changes in building codes needed; quality of construction material is important
Skills in unorganized construction sector	Mason training and certification	Hunnarshala and VRTI-Vivekananda Research and Training Institute (local NGOs) were involved; masons are losing skills
Training on relief and emergency response procedures	For government staff organized by GIDM	Largely supply-driven
Awareness programs	For school children by DEOC	More programs are needed for every school

6.5.2.2 Relationship Between Forms of Capacity Development and Perceived Capacity Deficit

The field research suggests that the relationship between capacity development efforts and capacity deficit perception varied with stakeholders and with time. For the capacity development phase just after the earthquake, government and community groups took various actions based on their perception of capacity deficits. However, in the long run, community groups and citizens started scaling back or completely abandoning capacity development activities due to a lack of funds from the government and donors or a sense of false security from improvements in the house construction. The majority of interview respondents said that the government was aggressive in the first three to four years from the start of GEERP, but that slowed down in the later years. The discussion below and Table 6.11 summarize the findings.

Government and Donors

The basic design of the GEERP was based on assessments conducted by the government and donors, which identified capacity deficits in the areas of a dedicated organization for disaster risk management, technical capacity development, and awareness. After the project was over in 2009, donors' completion report raised the issues of sustainability of the program including continued interest and engagement of local government bodies and community members, accountability, and code compliance (World Bank 2009). However, from the donors' side, the capacity development effort was over, and it was now the state government's responsibility to ensure sustainability. The state government, on the other hand, believes that sustaining the momentum generated through the GEERP requires better code compliance, accountability, and closer links with local governments. However, they are not planning any new capacity development efforts in those directions.

Community Groups

Just after the earthquake, interest groups such as the builders association and religious- and caste-based groups perceived the greatest deficits as immediate relief, relocation, rehabilitation, and reconstruction. Accordingly, several of these groups adopted villages and urban areas to help with relocation and in situ reconstruction. Even over a decade after the earthquake, the groups still believe that the capacity is low, but they are not taking any actions since they believe that their role involves actions needed during emergencies only and does not involve prevention. Major NGOs have mentioned a long list of capacity deficits that continue to hamper preparedness. However, their areas of focus are primarily training and emergency response and not disaster preparedness and prevention. They have undertaken small-scale capacity development efforts relative to the identified capacity deficits, depending upon the availability of funding. Their impact remains small and unsustainable. As one of the interview respondents from an NGO said:

> We work mostly in villages and slums. We are not directly doing anything on awareness other than the mason training. We work with another NGO on slum upgradation. This is also very limited (up to 75 houses). While the other NGO looks after the technical details on building construction, we are involved in beneficiary selection and social development such as counseling. Even after us telling masons about how to construct, they still work using the old ways.

Similar to respondents from the NGOs, the majority of citizens surveyed have a long list of capacity deficits that remain through the years. Just after the earthquake, they took measures such as owner-driven reconstruction and relocation, using governmental incentives to reduce their risk. Although a majority of the surveyed citizens are aware of current risks, prevention activities are not a big priority for them. Their capacity development efforts relate to their economic profile and the potential tradeoffs between costs of prevention activities and the economic benefits. For example, some vegetable sellers in Bhuj even now sit beside a potentially dangerous historic building with huge cracks (see Fig. 6.11). Even though they know that the building might fall down anytime, they are willing to sit up against its walls since it is the only place in the old city market that they have secured for themselves, and that place is directly tied to their daily earnings. Similarly, some citizens belonging to low-income groups continue to live in temporary shelters even today. Their homes are susceptible to damages during earthquakes, but they are more economical compared to constructing new safer houses.

6.5.3 Mechanisms for Formulating and Monitoring Capacity Development Interventions

This section examines the mechanisms that were used by the donors and the government to formulate intervention designs, implement actions, and monitor progress. The mechanisms are similar to typical donor-driven approaches in post-disaster

Fig. 6.11 Vegetable sellers in Bhuj sitting next to a damaged building (Source: Author 2012)

scenarios that reflect top-down planning, best practices adaptation, rapid creations of institutions, minimal civic participation in project design, limited integration with bottom-up processes, overreliance on quantitative indicators to measure progress, and setting an unrealistic project timeline.

6.5.3.1 Top-Down Planning

The field study shows that top-down planning drives much of the project planning and the designation of project intervention areas. The assessments undertaken by the state government and donors (Joint Assessment Report) formed the basis of assessing capacity deficit areas, project intervention areas, funding, and timing. The state government's and donors' past experiences and learning from similar situations also played a key role in shaping the response. Since disasters pose a very serious humanitarian situation, funding was probably well above the state government's normal allocations. A top-down response was probably necessary to support the massive post-earthquake relief and recovery response, since much of the infrastructure and community assets were lost. The top-down nature of the interventions, however, provided little design flexibility and limited support for the environment or implementation:

- *Little design flexibility*: The project design did not change much over the course of project implementation. There were some changes after there was community opposition to contractor-driven reconstruction approaches and dispute over compensation packages, but, overall, the design remained the same.
- *Little understanding and support for the enabling environment:* The project design document did not elaborate much on the environment within which the project was taking place. Indirect benefits for the enabling environment may have been created during the course of project, but no intentional support was provided.
- *Reliance on actions rather than implementation:* The project design focused more on actions than on implementation arrangements and involvement of communities, as was evident in issues faced during implementation such as opposition from communities regarding relocation, contractor-driven housing reconstruction, and packages.

6.5.3.2 Best Practices Adoption

All three capacity development forms—formation of new agencies, reconstruction, and training—are viewed as best practices and are commonly adopted by donors (Kenny 2009). Two innovative best practices were included in project design: focus on disaster risk management and earthquake-resilient reconstruction designs. The long-term disaster risk management component of the project emerged from the World Bank's earlier experience in handling post-disaster reconstruction projects (World Bank 2005). Both of the best practices provided innovation; however, there were questions regarding the sustainability of earthquake-resistant reconstruction designs,[2] especially in rural areas where institutional capacity to implement and monitor waned after the project was over. Limited community awareness, involve-

[2] The transfer of knowledge on best practices in construction proved effective at sustaining and improving the application of seismic resistance techniques during the reconstruction process. However, field assessments suggest that some of these best practices have subsequently been

ment, and non-similarity with traditional Gujarati houses have been a key factor in not being able to sustain the design (Sanderson and Sharma 2008).

6.5.3.3 Rapid Creation of Institutions

The GSDMA was created within a few weeks to ensure coordination between different government departments, line ministries, donors, and NGOs. Many new administrative agencies and authorities were also created: the Gujarat Urban Development Company (to manage implementation of reconstruction projects) and four area development authorities (ADAs) were created to develop and implement development plans in four severely affected urban areas. These institutional structures were developed quickly to support disaster response, with no exit strategy, meaning with no understanding of how they would continue after the response was over. While GSDMA continues as the state's disaster management agency—with the mandate of preparing the state to deal with disasters—ADAs had to hand over authority to the municipalities. Other institutions created include the Gujarat Institute of Disaster Management, which was created in 2004 to develop human resource capacity in disaster risk management and the Seismological Research Institute. We can see elements of "wishful thinking" and "premature load bearing" through the example of GSDMA. However, unlike what Pritchett and Weijer (2010) proposed, i.e., such rapid creation can hollow government's capacity, the role of GSDMA has been recognized by the international community (with the conferring of the UN Sasakawa award), national and state governments, and local NGO groups. A leader of a very powerful local NGO group said in an interview that the creation of GSDMA was very critical to the largely successful response. He appreciated the cooperative approach undertaken by GSDMA to work with NGOs. Arguably, the entrepreneurial culture of Gujarat, the proactive government, the grafting of seasoned IAS officers, and the committee consisting of Gujarat's eminent academicians and politicians to oversee planning, as well as good media attention, may have been behind the success of GSDMA.

6.5.3.4 Minimal Civic Participation in Project Design

According to World Bank (2009), community involvement included "...information and communication activities, establishment of village level institutions, strengthening local governments, gender sensitization and community based disaster awareness and preparedness programs." Measurable indicators included (i) the number of self-help groups constituted in villages and (ii) the number of disaster management committees formed to reduce the vulnerability of communities against future emergencies and their capacity building through stakeholder participation (see Table 6.9). The field study clearly shows that disaster management committees

---

ignored, particularly in rural areas, due to the waning institutional intermediation (World Bank 2009).

Table 6.9 Relationship between capacity deficit and capacity development efforts

Perception of:	Capacity deficit (before and just after 2001 earthquake)	Capacity development efforts (just after 2001 earthquake)	Current capacity deficit	Current capacity development efforts
Donors	Lack of a sustainable disaster management program (World Bank and Asian Development Bank 2001)	GEERP: New organizations and institutional structure, reconstruction, training/skill building/education/awareness	Sustainability of the program Accountability Code compliance	None planned
State/district administration	No DRM act/policy/structure/ training	GEERP: New organizations and institutional structure, reconstruction, training/skill building/education/awareness	Equipment acquisition is slow No employee rewards	Early warning systems Microzonation studies Building codes review
Municipality	No urban development plans or no updates No control on development and building specifications	Training provided by the GSDMA	Awareness programs Coordination Enforcement Community need to value life Warning systems	None planned
Development authorities	Did not exist	New authorities setup Reconstruction City planning	Enforcement power and mechanisms Staff and funding Citizens do not follow rules	Requested for more funding and staff from government
Private sector, interest groups	Massive relief, recovery, and rehabilitation need Coordination of efforts	Relief, recovery, relocation, and rehabilitation for all or selected community groups	Coordination is patchy Lack of clarity in roles and responsibilities leading to duplication or gaps	None
NGOs	Disaster management was not a big priority Coordination efforts started after the cyclone of 1999 by KNNA, collaboration of NGOs	Training Reconstruction, relief, and rehabilitation	Awareness programs Participatory approach Better coordination and enforcement Need to link preparedness with livelihood activities	None Focus is on slums and villages on poverty side, not on DRM

(continued)

Table 6.9 (continued)

Perception of:	Capacity deficit (before and just after 2001 earthquake)	Capacity development efforts (just after 2001 earthquake)	Current capacity deficit	Current capacity development efforts
CBOs	Disaster management concept was not in place	Reconstruction	No preparedness No alerts Awareness program need to be strengthened People are not taking responsibility	Awareness programs
Citizens	Need to recover Not aware of the need for proper property records, safe housing techniques, and do's and don'ts during emergency	Reconstruction Relocation	Illegal construction Risky buildings not yet removed No citizen committees Not able to strike balance between economic and safety considerations	Ensure that legal property records are secured Safer housing construction if economically possible

do not exist, at least in the study towns. Community involvement was more passive to increase awareness rather than to actually involve the community in designing capacity development interventions.

6.5.3.5 Limited Integration with Bottom-Up Processes

Local NGOs played a key role in the project. They organized communities and worked with them to push their demands and concerns to the government. The government was not very concerned with how communities were integrated into the informal processes that were undertaken during the earthquake response period. A powerful local NGO highlighted the following issues, which could have been avoided with greater community involvement:

- The difficulties resulting from pressures from influential locals to relocate villages, including the creation of various "ghost towns" that included Vondh that was built by the government of Maharashtra, which includes about 1,000 houses, or the township of 282 houses built outside Bhuj
- The dumping of debris in Pragsar lake, one of the three ancient lakes of Bhuj that was established hundreds of years ago as part of a system that recharged the sandstone aquifer
- Wastage in the livelihood programs due to poorly designed and inflexible support through toolkits, weaving looms, and farming kits
- Too much focus in urban areas on housing rehabilitation rather than support for the recovery of small businesses

6.5.3.6 Reliance on Quantitative Indicators to Measure Progress

Consistent with other World Bank projects, key indicators for achievement in each sector were defined in the beginning with measurable targets, particularly, for support for disaster management capacity building. This included enhancing the state's disaster management capacity through (i) setting up of regulatory and research institutions, (ii) preparing state level disaster management plans, (iii) setting up and functionalizing a statewide emergency communication network, and (iv) setting up and functionalizing statewide control rooms. The donor and government reports regularly monitored numerical project indicators (based on GSDMA's quarterly reports and the World Bank and ADB project completion reports). The donors have also accepted the project outcomes (in terms of achieving numerical targets) as "satisfactory" (see Table 6.10). The quantitative target does not necessarily indicate any sustainable capacity building of the government or the community.

Other than reporting on numerical targets to donors, the government of Gujarat also conducted two other assessments to monitor results. These are discussed below and are better measures of progress.

Table 6.10 World Bank-supported project target values and outcomes

Original target values (from approval documents)	Formally revised target values	Actual value achieved at completion or target years
Estimate date: 10/31/2005	*Revised date: 10/31/2008*	*Completion date: 03/31/2009*
Project indicator 1: Rehabilitation and reconstruction program in the earthquake-affected areas including restoration of houses, public buildings, and basic infrastructure in the roads and irrigation sectors completed		
House reconstruction: 135,000	House reconstruction: 125,000	Houses reconstructed: 125,781
House repair: 75,000	House repair: 42,000	Houses repaired: 41,751
Trained engineers: 1,000	Trained engineers: 1,000	Engineers trained: 678
Trained masons: 5,000	Trained masons: 5,000	Masons trained: 3,832
Dam rehabilitation: 222	Dam rehabilitation: 225	Dams rehabilitated: 225
Public buildings reconstruction: 1,200	PB reconstruction: 200	PB reconstructed: 232
PB retrofitting: 500	PB retrofitting: 2000	PB retrofitted: 2,848
New buildings for GIDM, ISR	New buildings for GIDM, ISR	New buildings: ISR, GIDM (in progress)
Improve roads: N/A	Improved roads: 800 km	Roads improved: 870 km
Project indicator 2: Institutional framework to allow better disaster mitigation and risk management for future natural disasters developed		
Reduce community vulnerability: Constitute 500 self-help groups	Reduce community vulnerability: Form DM Committees	Reduced community vulnerability: DM committees formed in 10,289 villages
Conduct 100 *gramsabhas*		
Set up institutional framework: Legislate DM Act	Set up institutional framework: Legislate DM Act	Set up institutional framework: DM Act legislated
Improve DM capacity: Statewide emergency communication network	Improve DM capacity: Statewide emergency communication network	Improved DM capacity: Expert committee on statewide communication was constituted

Source: Asian Development Bank (2008), World Bank (2009)

(i) Baseline Monitoring and Evaluation Reporting (KPMG 2005): The Quality of Life Index[3] analysis done as part of the benefit monitoring evaluation suggests that the quality of social and built environments in the disaster-affected communities not only has been restored to pre-disaster levels but also has shown a progressive improvement during the post-disaster period.

(ii) Social and Poverty Impact Assessment (CEPT University 2009; Sharma 2009): Focus group discussions and three surveys were conducted in affected towns and villages. The studies found overall positive impacts but brought forward the

[3]The index, normalized for baseline values to 1, is based on the following indicators: access to house/permanent house, size of house, and basic amenities in house—separate kitchen and toilet; access to primary school, basic amenities in primary school; access to functional health facility; access to surfaced road; adequate drinking water; access to piped water supply/public stand post; and access to electricity. The surveys were largely conducted in rural areas.

need to enhance community networks and undertake community skill building and employment generation programs. At the same time, the studies brought to light some dissatisfaction with the earthquake recovery project in terms of corruption and ethnic issues. The study notes "the momentum has not been sustained to a stage where community development and management issues could be addressed. The flexibility of the Government has been misused for ulterior motives" (p. 79).

6.5.3.7 Unrealistic Project Timelines

The project was initially conceived for three years, which was a very short timeline to achieve desired results. As mentioned in the Asian Development Bank Implementation Completion Report (2008):

> Implementing a large-scale reconstruction program within the three years originally provided was not practical. Due to the scale and spatial spread of destruction caused by the earthquake, the consultation and planning process took nearly two years, leaving little time for actual reconstruction, which took an additional three years. (p. 2)

In fact, the World Bank-supported components took seven years to complete. The World Bank's project timeline was extended three times with the original project completion date extended from 2005 to 2007, 2008, and then 2009 (World Bank 2009). Both the World Bank and ADB reports point to issues related to approval delays and less experience of GSDMA in dealing with procurement procedures. It is interesting to note that GSDMA officials, on the other hand, complained about the cumbersome World Bank procurement procedures. Various project fixes related to adjusting statistical targets depending upon implementation, revising the timeline, and resource readjustment were undertaken.

6.5.3.8 Heroic Start After the 2001 Earthquake; Dampening Momentum a Decade Later

In summary, the state government, donors, and the community (NGOs, CBOs, businesses, and citizens) clearly linked the capacity deficit perception to the capacity development efforts just after the earthquake. Capacity development just after the earthquake followed three broad forms under the GEERP: setting up of new institutions, reconstruction, and training and raising awareness. One decade after the 2001 earthquake, most of the stakeholders believe that disaster risk is still high and capacity low, but they are either not taking any actions or taking minimal actions to reduce perceived capacity deficits. These deficits include damaged buildings yet to be retrofitted, poor building construction quality, illegal construction, lack of enforcement, lack of funds, lack of clarity on roles and responsibilities among government agencies, lack of awareness, and lack of collaboration among government, NGOs, and citizens.

Just after the 2001 earthquake, both top-down and bottom-up interventions were taken up in study towns, with clear links between identified capacity deficits and disaster risk. This is as per the new conceptual model. However, a decade later, such links are not visible, showing that the capacity development momentum has slowed down. This section also shows that design, implementation, and monitoring of capacity development interventions under the GEERP followed typical donor-driven capacity development approaches including top-down planning, best practices adoption, rapid creation of institutions, minimal civic participation in project design, limited integration with bottom-up processes, overreliance on quantitative indicators to measure progress, and setting an unrealistic project timeline.

The next two sections discuss whether the GEERP resulted in effective and sustainable capacity development.

6.5.4 Capacity Development Under the GEERP

Did the GEERP undertaken by the government of Gujarat result in capacity development? This question focuses on whether effective capacity development, as defined in the conceptual framework, occurred in the study towns after the 2001 earthquake. Unlike the numerical indicators used by the donors and the state government, as discussed in the earlier section, this section presents the effectiveness of capacity development by analyzing (i) state and local government capacity (consisting of policy, organization, implementation, and political changes) and (ii) community capacity (consisting of skills, coordination, cooperation, leadership, and inclusion). The findings in this section are derived through interviews and surveys.

The three study towns have well-established disaster risk management activities as discussed below. The overview is based on a review of available documents as well as interviews with state, district, and local government officers.

(i) *Preparedness*

- *Disaster warning infrastructure*: exists for cyclones and floods at the district level; seismic monitoring and alert system is being developed.
- *Contingency plans:* exist for all three study towns.
- *Coordination mechanism*: no citizen committees exist, but the DEOC coordinates contingency planning activities with major stakeholders.
- *Training*: currently offered by GIDM to staff of municipalities and ADAs.
- *Public awareness on disasters*: the DEOC undertakes limited awareness programs for selected schools.

(ii) *Risk reduction or mitigation*

- *Risk assessment mapping and modeling*: being developed by the GSDMA; state level risk maps are available to the public.
- *Risk-based land-use planning*: carried out after 2001 in Bhuj and Bhachau.
- *Building codes*: exist but are old; review is in process.

(iii) *Response, recovery, and reconstruction*

- *Institutional structure:* exists at national, state, and district levels with dedicated budget and staff

Does this mean that different levels of governments and local communities have required capacity? To find out, government and community capacity was assessed through PeDJoLA model.

6.5.4.1 State, District, and Local Government Capacity

Policy Capacity

Policy or institutional capacity is the ability to uphold authoritative and effective rules of the game, which comes from enabling laws, policies, and programs. The Gujarat government was swift to establish legislation regarding disaster management and set up the GSDMA to implement the law. A DEOC official talked positively about the DRM plans at the district and local levels:

> Police, irrigation, and other departments make departmental disaster management plans and we collate them into a district plan. The plans include potential impacts, response required, who is in charge, who has the equipments, who the response team members are, etc. The Irrigation Department will report the number of dams, overflow details, and potential effects. We coordinate review meetings to follow up and monitor.

However, the interviews suggest that although the framework, policies, and programs exist, the community is mostly unaware of these and that they are not effectively working in practice. Most NGO respondents mentioned that they do not see effective policies or plans. As one NGO respondent commented:

> There is a state level policy for response but nothing in terms of preparedness. Or at least I am not aware of any transparent policy. The stress was on rehabilitation and reconstruction in the first 2–3 years. After that, there is nothing much happening. Nowadays, the only thing government agencies do is making colorful reports.

A retired government officer interviewed corroborated this:

> I have heard of the State Act and GSDMA, but they are not well publicized and visible.

Another NGO respondent commented on the effectiveness of such policies and plans:

> State and local government guidelines are not filtering to the ground.

ADAs consider their city development plan and building regulations as DRM plans and suggest that they are effective, though there are issues with enforcement. As the ADA representative in Bhuj put it:

> BHADA was established after the earthquake. Consultants prepared Development Plan, Development Control Regulations, and Town Planning Schemes [a method of land development]. The Town Planning Officer was appointed directly from Gandhinagar [state capital]

> under a CEO. The Chairman was appointed later. There are four engineers here. Staff goes for workshops and trainings. However, if citizens do not follow rules, nothing can be done.

GSDMA's senior officer at the state level also accepted it as an area that needs attention and suggested that this is an issue with the leadership at the local level:

> The structure is there but implementation is debatable as it depends a lot on individuals and their leadership skills.

The surveys also supported the opinions expressed in the interviews. Two questions were asked in the survey of the government staff to understand whether DRM policies are considered in planning disaster prevention plans and programs and if effective DRM plans exist.

A majority (69 %) of the local government respondents across different agencies agreed that DRM policies are considered when creating disaster prevention plans and programs in their agencies. However, only half of the interview respondents, mostly municipal staff, think that effective DRM plans exist in the surveyed towns. The difference could be attributed to the fact that municipalities have the least role in disaster-related plans and programs in the current structure of roles and responsibilities. All surveyed towns have high agreement on whether DRM policies are considered in developing plans. However, contrary to the other two towns, government officers in Bhuj do not believe that their DRM plans are effective.

In summary, there are enabling laws, policies, and programs at the state level to improve disaster risk management. More steps are needed to ensure effectiveness of plans.

Organizational Capacity

Organizational capacity relates to the state and local government's internal organization and management style related to the structure and distribution of functions, planning, decision-making, control and evaluation functions, and information gathering, processing, and distribution. The interviews suggest that there is lack of clarity of roles and functions as well as issues with coordinating activities.

DEOC staff interviewed state that their mandate is clear and they clarify the roles of all agencies while coordinating DRM plans:

> We coordinate with other agencies in the district. All Class I and II officers come from all over the district for meetings. Taluka Mamlatdar, Deputy Mamlatdar, and TDO also come for meetings. We talk about roles and plan implementation.

However, municipalities think that their role is limited to disaster response. As a Mayor mentioned:

> Municipality does not deal with building permits or other plans. We are responsible for disaster response and relief work.

On the other hand, ADAs consider their role to be solely focused on development and building construction:

> Our work does not overlap with most other agencies. We make land-use/infrastructure plans and approve construction. Once an area is developed according to standards, we hand over the maintenance responsibility to the municipality.

NGOs also point to this lack of coordination during non-disaster situations:

> I don't see any coordinated activities among government agencies. There was better coordination right after the earthquake. It is no more there.

As a retired government officer pointed out, the enforcement is lacking and the reason could be the coordination issues:

> ADA is supposed to ensure plan enforcement and quality control. In this town itself, there are about 30 buildings that ADA has identified as unsafe and in need of demolition. However, neither they nor the municipality has taken any action in more than 6–7 years.

However, the survey of the government officers shows that there is a clear administrative structure, better coordination, and reasonable information sharing with citizens. Three questions were asked to understand capacity related to administrative structure, coordination, and information sharing. Overall, a majority of the government respondents agreed that agencies have a clear role (69 %), they coordinate well (82 %), and information is shared well with the communities (53 %). While all other agencies agreed about the coordination, ADAs had less agreement (only 33 %). The interviews suggest that this could be due to perception of ADAs that they work independently of other agencies. There was less agreement on information sharing in Bhuj (38 % agree) compared to Bhachau (86 %) and Mandvi (75 %). In summary, government staff agree that organizational capacity exists, and the three local agencies believe they have clear roles before, during, and after disaster. Responses indicate that a disaster response protocol exists that describes how agencies will coordinate. While information is shared with the community, it seems low in Bhuj.

Implementation Capacity

Implementation capacity is the ability to carry out decisions and enforce rules, within the public sector itself and in the society. The interviews revealed deficiencies in resources such as staff and funds, issues with enforcement, corruption, and attitudes of citizens.

Municipalities think that their role is limited to response and passed the responsibility to ADAs. One Mayor mentioned:

> Municipality does not deal with building permits and construction these days. It is done by ADA. We are responsible for relief work only.

It seems that there are problems with both clarity in responsibility and resources for enforcement. This is clear from what one ADA representative stated:

> A registered engineer must sign the building plan. Our supervisor goes and checks the completed building and issues a completion certificate and use permit. People illegally

construct after the issuance of the certificate. There is no system to stop that. If someone complains, we check them and provide notice and demolish. But that happens in only very few cases.

As one NGO representative pointed out, corruption could be a major factor in weak enforcement:

The builder lobby, government officials, and politicians are together in corruption.

As many respondents mentioned, there are serious issues with the attitudes of people. One municipal councilor said:

People bribe officers to do illegal construction. They don't understand that this can take their own life. Attitude change is needed.

It is also interesting to note that both government officers and a news reporter in Bhachau agreed that people in their town follow regulations more than other towns in the district. One of the municipal officers in Bhachau mentioned that the people complain to authorities if they see illegal construction. The news reporter said:

People here no more construct concrete/brick overhead water tanks on top of the buildings, which were a major cause of deaths and damages during the 2001 earthquake.

In the questionnaire survey of government officers, they were asked if building codes are enforced well to understand this important aspect of government capacity. Overall, there was nearly an equal split of government respondents on building code enforcement (50 % disagreed, 44 % agreed, and 6 % had no opinion). While all of DEOC respondents disagreed that codes are enforced well, ADAs completely agreed and municipalities were in the middle (47 % agree). ADAs have the mandate to enforce building codes, and they were expected to agree with the statement (83 % strongly agreed and 17 % agreed). Respondents from Bhuj showed highest disagreement (57 %), while those from Bhachau were at the other extreme (72 % agreed). This supports the responses from people outside government in Bhuj and Bhachau.

In summary, there is a lack of technical staff and financial resources to carry out the implementation and enforcement of functions assigned to different agencies. The nexus among officers, politicians, and builders also leads to illegal construction that increases the disaster risk. The lack of communities' will to comply with the building codes also impairs the local government's capacity to implement and enforce plans.

Technical Capacity

Technical capacity relates to the quality of staff (skills, knowledge, and experience), how they are being used, and whether enough technological, financial, and motivational support is available for them to perform effectively.

Experience of staff, financial support, and motivation were primarily examined through the interviews. All of the interview respondents pointed out that there are no real warning systems other than cyclone warnings. Though DEOC was happy with the number of staff and resources available, one NGO representative questioned their effectiveness:

> You know that there are a total of five full-time staff at the DEOC. I don't know how they can deal with the activities in the largest district in India!

ADA representatives spoke positively about their capacity to guide development:

> Area development authorities were established and they have the required capacity to deal with development. (ADA)

However, it should be noted that ADA in Bhachau mentioned that they face problems due to a lack of sufficient staff as well as the lack of a professional planner:

> We should have had a Chief Executive Officer and Town Planner full time here. But right now, they are in deputation from Bhuj. Our CEO sits in Collector office with additional charges and Junior Town Planner sits in Anjar. How are we supposed to take quick and efficient decisions?

The Mayor of a municipality gave a clear picture about the situation in municipal offices. He stressed that there is a lack of technically sound and motivated staff. He also pointed to the absence of early warning systems.

> The local body even now doesn't have any new staff. Staff structure has the same make-up from 1952, when the city was very small. Today, the city has grown much bigger. But we still run with the same old number of staff. We are not allowed to keep more than 20 % permanent staff. We have cut down on 70 % new appointments. The rest, we run with temporary staff. Rs. 3500/month [USD 70] is the salary for them. They lack awareness and are not fully engaged or committed. They are appointed by recommendation from old staff and councilors. They are not skilled too. They look for permanent jobs elsewhere or do other business that becomes their primary objective. Once their business becomes successful, they leave the job. Therefore, there is no continuity too...... There is no early warning system. I keep hearing for many years that it will come.

Surveys of government staff gave a sense of their personal perceptions. Two questions were asked in the survey to examine technical knowledge and skills among the government staff and if effective disaster warning systems exist (see Table 6.11). A majority of government respondents agreed that staff have knowledge and skills (85 %) and that an effective warning system exists (72 %). While respondents from DEOC and ADAs agreed fully with the two statements, the

Table 6.11 Technical capacity perception of government staff

		Percent	Number of responses
Staff have knowledge and skills	Agree	85	27
	Disagree	6	2
	No opinion	9	3
Early warning systems exist	Agree	72	23
	Disagree	22	7
	No opinion	6	2

percentages of respondents from the municipalities who agreed were slightly lower (76 and 71 %, respectively). While most respondents from all three towns agree that the staff has knowledge and skills (70–75 % agree), the existence of a warning system is disputed. While 67 % of respondents in Bhuj agreed that the warning system exists, those from Bhachau and Mandvi were equally split between agreement and disagreement. This could be due to the fact that Bhuj is the administrative headquarters and has better access to information. In summary, government staff thinks that technical capacity exists in the study towns. The difference in the surveys and interviews may be due to the fact that the DEOC and ADAs are relatively new agencies that were set up under special circumstances and with dedicated staff and finding. Municipalities, on the other hand, have to do more work with fewer resources.

Political Capacity

Political capacity is the ability to mediate conflicts, respond to citizen demands, allow for representation of interests, and provide opportunities for effective community and political participation (Grindle 1996). In PeDJoLA model, this pertained to the awareness programs that provide information to the citizens, the interest and leadership from political leaders, and the level of community participation.

One of the main objectives of GSDMA is to raise awareness among citizens. The DEOC representative I interviewed showed me a number of posters and pamphlets that are sent out to municipalities and other government agencies for display and distribution. These included "do's and don'ts," DRM programs, and tsunami and industrial accident information. The DEOC representative mentioned:

> In addition, we carry out and encourage schools and other organizations to carry out competitions, debates, rallies, and drills on disaster-related topics. We also use local cable networks to send out seasonal prime-time advertisements. We also encourage observation of 'safety week', 'UN Disaster Reduction Day,' etc., in cities and villages. We collect reports including photos from organizers and include them in our bi-annual reports.

Representatives of NGOs and CBOs as well as a retired government officer were quick to point out that these were only cosmetic actions.

> There are no long-term actions to build awareness. Whatever they are doing is sporadic and coverage is very poor. Government should carry out more audio-visual displays, regular mock drills, and establishment/meeting of citizen committees. (Retired government official)

> These photos they showed you are from private schools where only high income and higher caste children go. How do you expect to have wide-spread awareness from these? (NGO representative)

While politicians interviewed showed interest in DRM activities, others were not so positive about political leadership. Comments from the retired government officer summarize this sentiment:

> Everyone says that they are interested in disaster management. Politicians or administrators do not take proactive actions other than the mandatory reporting. The new Mayor has

> started some activities on restoration of water channels, which is encouraging. However, that might change when someone else becomes Mayor.

As an NGO representative mentioned, political leadership might not be interested due to the lack of financial support for prevention activities:

> Every program with money is an avenue for corruption. There is not much money for preparedness, and therefore, not many politicians are interested to pursue this.

A political leader (councilor) opined that it is votes rather than sustainable change that attract local politicians:

> We don't pay much attention whether it can make comprehensive change. We do things only to the level where it brings votes.

In the survey of government officers, four questions were asked to examine capacity related to the publicity and awareness raising, leadership, and community. While most of the government respondents agreed that DRM actions are known to the public (66 %), that many leaders are interested in DRM (60 %), and that people are consulted in prevention planning (91 %), they largely disagreed with the statement that there was a change in DRM focus with leadership change (59 % disagree). While representatives from the DEOC and ADAs agreed that DRM actions are known to the public, municipal respondents were split in their opinion (48 % agreed and 38 % disagreed). However, all the agencies agreed that they consult with people in their prevention planning. While representatives from the DEOC and the municipalities agreed that many leaders are interested in DRM, ADA respondents disagreed. Representatives from the DEOC and the municipalities disagreed with the statement that there was a change in DRM focus with leadership change, but the majority of ADAs did not offer in opinion. When I compare the towns, it seems that both Bhuj and Bhachau agree that DRM actions are known to the public, but Mandvi disagrees. An opposite reaction is seen regarding the leadership change. All towns agree that there are many leaders showing interest in DRM and that they consult with the people during plan preparation.

In summary, political capacity related to disaster risk management exists in Bhuj and Bhachau. Mandvi, on the other hand, did not face large-scale destruction in 2001 compared to the other towns, and although leaders are generally interested in DRM, they do not seem to give the same level of attention to the topic. Even where political capacity exists, the leaders are not showing strong support.

Analysis of the Findings from the Interviews and Government Staff Surveys

Overall, government staff agreed that institutional, organizational, technical, and political capacities exist in local government agencies. The majority of them, however, also agreed that enforcement capacity is weak. Agency-wise, respondents from ADAs differed in their opinion from other agencies on questions related to organizational, implementation, and political capacity. Regarding organizational capacity, two thirds of the employees of ADAs who took the survey had no opinion on the clarity of roles, coordination, and information sharing. Unlike DEOC and municipality staff, ADA staff believed that building codes are enforced well, which is their core mandate. Finally, regarding political capacity, the ADA staff disagreed with the majority view that many government leaders are interested in DRM. It is interesting to note that ADAs are now headed by political representatives, which was not the case when they were initially grafted into the system just after the earthquake. The municipalities had a slightly different response. Earlier, there was neither any dedicated staff nor any training. Municipal staff now receive biannual training on various topics such as firefighting and flood rescue. However, there is a severe shortage of staff to provide even essential services. This means that they are focusing more on essential services such as water supply and drainage than worrying about the possibility of a disaster. Disasters are only considered when they occur. Municipal staff consider DRM as DEOC's mandate now. As one municipality officer responded:

> Municipality did not have qualified engineers nor were there any restrictions/guidelines. Municipality did not have any staff for such labor-intensive/long work such as preparation/revision of city development plan. We still run with the same old number of staff but many of the positions from staff who retired were filled with temporary staff. We received and continue to receive training from GSDMA. OECs do it at the district and taluka levels. ADAs are also of great help. (Bhuj municipality)

Mandvi showed the most differences from the overall results followed by Bhachau and Bhuj. Government respondents from Mandvi had a difference of opinion regarding implementation, technical, and political capacities. Their opinion was equally divided (with 50 % agreeing and 50 % disagreeing) on enforcement of building codes and the existence of a warning system. They disagreed with the statement that DRM actions are known to public and largely agreed with the statement that there was a change in DRM focus after a change in the leadership. Responses from government staff of Bhachau differed only on implementation capacity, where a majority of respondents agreed that building codes are enforced well. Responses from government staff of Bhuj also differed only on one statement related to the institutional capacity, where a majority of respondents disagreed that effective DRM plans exist. These findings are summarized in Table 6.12 below.

6.5.4.2 Community Capacity

Community capacity is understood through citizens' skills and resources, coordination, cooperation, leadership, and inclusiveness. To examine community capacity, many questions were included in the interviews. In addition, community capacity was further explored through a questionnaire survey of the residents in Bhuj from different geographic areas and with different income status. The findings are discussed below.

Table 6.12 Summary of findings from government staff survey and interviews

Government capacity	Overall	Agencies			Towns		
		DEOC	Municipalities	ADAs	Bhuj	Bhachau	Mandvi
Institutional—enabling policy	Yes	Yes	Yes	Yes	Partly yes	Yes	Yes
Institutional—programs and plans	Partly yes	Partly yes	Partly yes	Partly yes	No	Yes	Partly yes
Organizational—administration structure	Yes	Yes	Partly yes	Partly yes	Yes	Yes	Yes
Organizational—coordination	Yes	Yes	Yes	Partly yes	Yes	Yes	Yes
Organizational—information sharing	Partly yes	Yes	Partly yes	Partly yes	Partly yes	Yes	Yes
Implementation—enforcement	No	No	Partly No	Yes	No	Yes	Partly yes
Technical—skills	Yes	Yes	Yes	Yes	Yes	Yes	Yes
Technical—infrastructure	Yes	Yes	Yes	Yes	Yes	Partly yes	Partly yes
Political—publicity	Yes	Yes	Yes	Yes	Yes	Yes	No
Political—leadership (many leaders)	Yes	Partly yes	Partly yes	No	Yes	Yes	Partly yes
Political—leadership (change in DRM)	No	Partly no	Partly no	No	No	Partly no	Yes
Political—community participation	Yes	Yes	Yes	Yes	Yes	Yes	Yes

Skills and Resources

Skills and resources relate to the ability of the community to have technical skills in developing a safer house, information resources about disasters and disaster management, and financial means to invest in a safe location before and/or during a disaster. NGO representatives were of the opinion that masons were given training and general awareness programs were carried out on safe building construction during the reconstruction phase but not anymore. One NGO representative said:

> Abhiyan, Hunnarshala, and Unnati [NGOs] trained masons to integrate traditional and modern technologies in building construction. However, the number of people trained was too small compared to the total population. Also, there are no follow-up activities.

Illegal construction by the residents (knowing that it is dangerous) was also an issue raised by many. However, it was the unanimous opinion by respondents in Bhachau that there is very little illegal construction in Bhachau. A respondent from Bhachau municipality mentioned:

> People are aware of the importance of building construction techniques and demand for better construction. Therefore, masons and builders also pay more attention.

Table 6.13 Skills and resources: perception of residents of Bhuj

		Percent	Number of responses
Know safe building techniques	Agree	54	27
	Disagree	44	22
	No opinion	2	1
Receive information from government	Agree	48	24
	Disagree	52	26
	No opinion	0	0

In the survey, the community respondents were asked if they know safe building techniques and if they receive information on disasters and disaster management from the government (see Table 6.13). The survey respondents were split on their opinion (54 % agreeing and 44 % disagreeing) about knowledge of safe building techniques. The knowledge of safe building techniques was high in locations where reconstruction activities occurred (mostly old suburbs and relocation sites) and low in other areas, especially in areas with low-income groups. There was slightly less agreement among women respondents about the knowledge of safe building techniques compared to men. This could be attributed to the prevailing social system where women play less of a role in building construction and related decision-making and so are less exposed to these issues. Financial resources were not included in the survey as everyone affected usually gets financial compensation from the government if there is loss or damage. The respondents were divided on the statement that they receive information from the government, with 52 % disagreeing and 48 % agreeing. All respondents from slums, old city, and new suburbs disagreed that they receive DRM-related information from the government. On the other extreme, a majority of residents from the old suburb (80 %) and relocation sites (90 %) agreed with the statement. It was also noted that the access to information went down as the income levels went down. Females had less access to information compared to their male counterparts. This may be due to the fact that women have marginally less access to information (e.g., newspapers) or literacy than men.

In summary, the citizens have divided opinions on their capacity associated with skills and resources. Generally, the better the income levels, the better the knowledge on safe building techniques and access to information.

Coordination

Coordination relates to the ability to work together effectively without duplication. From the DRM point of view, it pertains to the ability to work together before a disaster to clarify roles and responsibilities (citizen committees/groups), the ability to supervise and monitor activities (citizen committee/group meeting regularly), having activities to be responsive to the community aspirations, and the ability to resolve conflicts. In general, citizen committees do not exist functionally. Many interviewees were even surprised to hear about citizen committees. The response below from an NGO representative is the most common response I received:

> We don't know anything about citizen committees. What are they?

However, DEOC was emphatic in stating that they have established committees. One of the respondents from interest groups clarified that these committees are established as a response team rather than a group to gather people's aspirations and bring that into plans/programs:

> There is a list of who all can swim, who all have tractors, who has shelter/food facility (Swaminarayan Trust, etc.) This is more for response than prevention activities. There is no coordination of activities too. There is no follow-up of what happened, what didn't. There are no regular meetings or even update of the list. They call up a meeting every six months but attendance is very low.

A member of a citizen group that was very active right after the earthquake in helping with public consultation on rehabilitation and reconstruction activities probably summed it up best:

> For namesake. Government did some training programs in the beginning. But there is no continuous action. No one knows that there is something like this.

The survey of citizens corroborates this. Almost all of the respondents, irrespective of their location, economic status, or gender, disagreed on the existence of citizen committees, that activities are responsive to aspirations, and that conflicts are resolved by consensus (see Table 6.14). A small percentage of people from relocation sites who agreed to the existence of such committees also mentioned that these committees do not meet regularly.

Therefore, coordination capacity seems to be low in the community. It is slightly higher for high-income group respondents and relatively lower for low-income groups, slum dwellers, and women respondents.

Table 6.14 Coordination capacity perception of residents of Bhuj

		Percent	Number of responses
Citizen committees exist	Agree	2	1
	Disagree	92	46
	No opinion	6	3
Activities are responsive to aspirations	Agree	2	1
	Disagree	94	47
	No opinion	4	2
Conflicts are resolved by consensus	Agree	4	2
	Disagree	92	46
	No opinion	4	2

Table 6.15 Cooperation capacity perception of residents of Bhuj

		Percent	Number of responses
People cooperate during disaster	Agree	100	50
	Disagree	0	0
People volunteer during disaster	Agree	98	49
	Disagree	2	1
Business/NGOs help during disaster	Agree	96	48
	Disagree	2	1
	No opinion	2	1

Cooperation

Cooperation is the ability to help each other during disasters. Interviews give the feeling that, generally, people cooperate irrespective of their religious/caste beliefs and income status. Everyone agreed that people cooperate and volunteer during emergencies. Many respondents pointed out that NGOs have a large presence in the district and are committed.

This is measured in the citizen survey through three questions on people's cooperation, volunteering, and help provided by businesses and NGOs (see Table 6.15). Community respondents overwhelmingly agreed that people cooperate and volunteer and that NGOs and businesses help during disasters. This shows strong ties and acknowledgement of help received during the 2001 earthquake. The results did not change considerably when analyzed by location. Old city residents showed slightly stronger agreement with all the three statements compared to respondents of other locations. This might be due to the long-standing ties among the residents. Minor disagreement (10 %) from new suburbs about people volunteering might be due to the fact that these are new residents without many ties to each other. The results showed some differences among the different income groups with higher-income groups disagreeing that there is cooperation and that people volunteer during times of disaster. The results did not vary much with gender.

In summary, the community has a strong ability to cooperate, although the new residents may not have developed strong ties and trust among community groups.

Leadership

Leadership is the ability to bring positive change in the community. This leadership is expected to come from political leaders, religious/caste leaders, and elder citizens with influence in the community. Most interview respondents said that there is able leadership within communities. However, they were not aware of DRM-related activities that can bring change in the community. The following comment from an NGO representative summarizes this thought:

> Leaders are plenty. They need more awareness and support to make long-term changes.

In the survey, Bhuj residents were asked if the focus on DRM changed for good after a change in leadership (see Table 6.16). A majority of respondents disagreed

Table 6.16 Leadership capacity perception of residents of Bhuj

		Percent	Number of responses
DRM focus changed for good after leadership change	Agree	16	8
	Disagree	80	40
	No opinion	4	2

(80 %) that there was any change in focus from the previous situation where no one paid much attention to DRM. This shows that new political leaders have not really provided strong support and leadership in regard to the DRM agenda. It may also mean that DRM is not a priority agenda for new leaders or that the current DRM program is so well established that the new leaders do not want to make any changes to it. While there is some difference in opinion among residents in the new suburbs (40 % agreed, 40 % disagreed), about one third of high-income group respondents strongly agreed that there has been a change in DRM focus after changes in leadership. Both males and females disagreed about the focus change.

Inclusion

Inclusion relates to the ability of the society to include weaker sections of the society such as lower economic/caste groups and women. Since the society in India is divided into caste groups and low caste groups are usually vulnerable, their inclusion was posed as one of the questions. Women's inclusion was also included since they are also considered weak in the society. Overall, inclusion was understood through the sense of ownership in DRM activities.

In general, interview respondents were positive about the inclusion of caste groups and women in community activities. Many noted that even when the rest of the state went through communal unrest in the early 2000s, Kutch district was calm and inclusive of all religious and caste minorities. One NGO representative stated that:

> Religious and caste differences are not reflected in daily activities here.

This shows high community capacity. However, when it comes to dealing with the government, powerful groups have more influence. This idea was suggested by a representative from a citizen group interviewed:

> Usually, powerful groups hijack agenda, particularly with the government.

In the survey of residents, most agreed that voices of caste groups (96 %) and women (86 %) are heard. There also appears to be strong ownership of DRM projects (86 % agreed) (see Table 6.17). However, it is clear that the slum residents and low-income groups feel that the female voices are not heard all the time (40 % and 36 % disagreement by respective groups). Respondents from old suburb and high-income groups did not feel as strong ownership as other groups. The results did not change with gender.

In summary, the communities have a strong ability to include different caste groups and women, and there is strong ownership of DRM projects. It seems, however, that

Table 6.17 Inclusion capacity perception of residents of Bhuj

		Percent	Number of responses
Caste groups' voices are heard	Agree	96	48
	Disagree	4	2
Women's voices are heard	Agree	86	43
	Disagree	12	6
	No opinion	2	1
Strong sense of ownership in DRM projects	Agree	86	43
	Disagree	12	6
	No opinion	2	1

there is still room for improvement to make groups feel ownership of activities that affect them directly. This may even point to the need for more awareness programs.

High Cooperation and Inclusion, Low Coordination, Skills, and Leadership

Overall, community respondents had mixed opinions regarding community capacity in disaster risk management. In particular, the community had divided opinion on their capacity with regard to skills and resources. Generally, the higher the income levels, the more they reported having knowledge about safe building techniques and access to information. In general, the coordination capacity seems to be low, although it is slightly higher for high-income group respondents and relatively lower for low-income groups, slum dwellers, and women. Communities have a strong ability to cooperate, which is an asset that can be built upon. Leadership capabilities are also very high in communities. The community has a strong ability to include different caste groups and women, and there is strong ownership in DRM projects. It seems, however, that there is still room for improvement to make groups feel ownership of activities that affect them directly. This could even point to the need for more awareness programs. Table 6.18 summarizes these findings.

6.5.4.3 Putting the Pieces Together: Increased Government Capacity, Comparatively Less Community Capacity

This section combines the findings from the interviews and surveys with government staff and the communities to develop a whole picture. Government staff thinks that the capacity for DRM has improved except in implementing building codes. While the state level government staff is more convinced that capacity has improved greatly, the local staff is relatively less convinced. This is especially true regarding the institutional (effective DRM plans), organizational (information sharing), and implementation (building codes) aspects of the capacity. Community leaders largely agree that capacity has increased. The NGOs, CBOs, and citizen groups agree that it is good to have such a government structure as the GSDMA but emphasize the need to sustain the intensity of activities and involve citizens (see Table 6.19 below for a summary).

Table 6.18 Overall summary of community capacity

Indicators	Overall	Location					Income group			Gender	
		Old city	Old suburb	New suburb	Relocation site	Slum	LIG	MIG	HIG	Male	Female
Skills and resources	Partly yes	No	Yes	Partly no	Yes	No	No	Partly yes	Yes	Yes	Partly yes
Coordination—committees	No	No	No	No	No	No	No	No	No	No	No
Coordination—conflicts	No	No	No	No	No	No	No	No	No	No	No
Coordination—responsive	No	No	No	No	No	No	No	No	No	No	No
Coordination—information	Partly no	No	Yes	Partly no	Yes	No	No	Partly yes	Yes	Partly no	Partly no
Cooperation—people	Yes	Yes	Yes	Yes	Yes	Yes	Yes	Yes	Yes	Yes	Yes
Cooperation—business	Yes	Yes	Yes	Yes	Yes	Yes	Yes	Yes	Yes	Yes	Yes
Cooperation—volunteer	Yes	Yes	Yes	Yes	Yes	Yes	Yes	Yes	Yes	Yes	Yes
Leadership—change	No	No	No	Partly no	No	No	No	No	Partly no	No	No
Inclusion—caste groups	Yes	Yes	Yes	Yes	Yes	Yes	Yes	Yes	Yes	Yes	Yes
Inclusion—women	Yes	Yes	Yes	Yes	Yes	Yes	Partly yes	Yes	Yes	Yes	Yes
Inclusion—ownership	Yes	Yes	Yes	Yes	Yes	Yes	Yes	Partly yes	Yes	Yes	Yes

Table 6.19 Opinion on government capacity

Government capacity	State government staff opinion/ information from secondary resources	Local government staff opinion	Community opinion
Institutional—enabling policy	DRM laws, DRM programs, and projects at national, state, and local levels exist	Partly yes	Yes
Institutional—effective programs and plans	Risk mapping and mitigation planning (ongoing), monitoring, DRM data collection, early warning systems ongoing	No (at local level)	No
Organizational—administrative structure	Clear roles and responsibility before, during, and after disasters	Yes	No
Organizational—coordination	Coordination plans or protocols between departments and agencies	Yes	No
Organizational—information sharing	Public awareness plans, training, and awareness programs	Partly yes	Partly no
Implementation—enforcement	Building codes enforcement	No	Partly no
Technical—skills	Trained staff, access, and motivation for training	Yes	No
Technical—infrastructure	In handling emergency—no early warning system yet, but response system exists	Yes	Partly yes
Political—publicity	Information available on DRM	Yes	Partly no (no or limited knowledge on safe buildings)
Political—leadership, many leaders	Many leaders interested in DRM	Yes	Yes
Political—leadership, change in DRM focus	Political leaders support or move the DRM agenda forward	No	No
Political—community participation	DRM and response planning	Yes	Yes

Table 6.20 Opinion on community capacity

Community capacity	Community opinion	Government staff opinion
Skills and resources	Partly no	No
Coordination	No	No
Coordination—government responsive	No	Yes
Coordination—information	Partly no	Yes
Cooperation	Yes	Yes (only during disaster)
Leadership—change	No	No
Inclusion	Yes	Yes
Inclusion—ownership	Yes	Yes

A majority of community respondents believe that their capacity is still low regarding DRM. While the abilities with respect to skills and resources and coordination are low, leadership, cooperation, and inclusion abilities are high. Interviews revealed that cooperation is also primarily during disasters only, not for prevention activities. All the interview respondents further point to the community's negative attitude of not paying attention to safe housing practices and for putting enough pressure on the government to perform well in reducing the risk (see Table 6.20 above for a summary).

As per the conceptual model, effective capacity at the local level consists of government capacity and community capacity. The field research indicates that the study towns have seen increased capacity in many government and community aspects. However, there is much improvement needed for them to be effective. The following comments could be considered as a summary of the capacities developed and their effectiveness:

> State and local government guidelines are not filtering to the ground. (NGO representative)

> Earlier, there was no focus on disaster. Now the structure is there. However, they need to move beyond day-to-day reporting to actual activities on ground, work with people. (Citizen group representative)

6.5.5 Sustainability of Capacity Development Efforts

Is the capacity development of local governments and communities, as achieved under GEERP, sustainable? This question focuses on the sustainability of the capacity development process after completion of the GEERP. This is primarily understood through how learning is taking place and how empowered local governments and communities are to carry out their responsibilities, reduce disaster risk, and increase capacity to deal with future disasters. This section relies heavily on interviews and surveys.

6.5.5.1 Learning

Learning denotes an internalization of lessons from past experiences and passing them on to the successors (both in government and community). It comprises (i) flexibility and openness to change, (ii) information sharing and flow, (iii) education programs, (iv) double-loop learning, and (v) innovation.

Flexibility and Openness to Change

Many interview respondents stressed the shortage of government staff as well as overload of temporary staff, which affects their ability to make changes.

> Municipality doesn't have any new staff. We work with temporary employees who are not committed enough. (Municipality representative)

Moreover, the top-down nature of administration provides limited decision-making power to the local staff who have to live with a more supply-driven way of working. This is evident, for example, in the decision-making occurring at the state level regarding the type of emergency vehicles to be delivered and the training provided for local staff without consulting local staff for their needs.

> We receive GSDMA grants and state budget for equipment and activities. They let us know about the training programs and we send staff. However, this is very top-down approach where State government decides what is needed. Many times, our requirements are different. (Municipality representative)

On the community side, NGOs, CBOs, and the private sector are either reliant on government and donor funding or are not interested in the risk prevention aspects.

> NGOs depend on government for funding and sometimes are forced to follow government/political directions. (Councilor)

The lack of NGO effort related to disaster prevention can also be attributed to the current focus of these NGOs. As one NGO representative said:

> We work on general capacity building, not specifically for disaster management. Our group was developed after the 1998 cyclone to coordinate NGO activities. But again, the stress was on relief and rescue. During and after the earthquake, we worked on rescue, relief, and rehabilitation. We do not have special modules on risk management as such now.

Thus, the interviews point to a limited flexibility and openness to change in the three study towns. However, the survey findings provide somewhat different picture. A majority of government respondents agreed in the survey that there is sufficient flexibility for them to take their own decisions (66 %). While 71 % of the government respondents from Bhuj and Bhachau agreed with the statement, 75 % from Mandvi disagreed. While all DEOC respondents and a majority of municipality respondents (67 %) agreed, only 33 % of ADA respondents agreed with the statement. Overall, 18 % of respondents did not record any opinion

Information Sharing and Flow

Some of the comments mentioned in earlier sections suggest that the information sharing among government agencies as well as with citizens and community groups is insufficient. The DEOC representative conveyed the impression that they are doing the best they can in spreading the message. However, the reach of their message, using traditional methods, is not sufficiently wide. A retired government officer criticized the approach of DEOC:

> DEOC messages in television say…'if you need information, contact us'…That is leaving the responsibility on citizens. That is not right. Government should take active steps to reach out to neighborhoods. I haven't seen that happening.

One of the Mayors interviewed suggested the need to go beyond the traditional methods:

> The physical programs and newspaper advertisements have limits. Very few people attend the programs and read print media. Even those who attend/read get the message only in long intervals. We need to utilize Facebook and other social media. Mouth-to-mouth can spread quickly and more efficiently. Once in the internet/social media, it will keep circulating, especially among the youth. This is important.

While a majority of government respondents to the survey think that information is shared well with the communities and people know about DRM actions, a majority of community respondents do not think that they receive DRM information from the government or other easily available sources. Lower-income group people, people living in old neighborhoods and slums, and women (who are the most vulnerable to the impacts of disasters) believed that they do not have easy access to information.

Interviews and survey findings in the earlier sections highlight that coordination among communities and with state and local governments is also limited. Thus, there are multiple reasons why enough flexibility and timely information sharing is not available in the study towns. Even if the information is available within the government, it does not appear to be reaching the communities to take timely actions.

Education Programs

One of the avenues for increasing the learning capacity of a society is to invest in education. Various aspects of disasters including prevention, life-saving methods, and emergency response could be introduced in the school and college curriculum to make the new generation aware. The GSDMA is working at the state level to bring about this change. A 3-year postgraduate diploma in disaster management was introduced in Lalan College, Kachchh University, Bhuj, several years ago. One of the senior staff in DEOC is a graduate of this program. The DEOC representative mentioned that disaster management is now a part of the school curriculum as well. This is in addition to the demonstration programs carried out in schools. However, representatives of NGOs and citizens interviewed mentioned that they are not aware of any such school programs. As mentioned in an earlier section, many also criticized the reach of such demonstration programs, stating it seems that these programs are only implemented in rich private schools in major towns.

> My son is in 7th grade. I pay close attention to his school work. I haven't seen or heard him talking about anything to do with disasters in his school work. (NGO representative)

Double-Loop Learning

Double-loop learning is put into practice when individuals or organizations modify their goal in the light of past experience for achieving that goal. It involves a learning process that leads to realization of what is working and what is not, what is achievable, and so on. It could result in modifying the original goal itself. In order to understand this, the surveys and interviews asked questions about the policy and organizational changes that occurred in response to the earthquake and the activities that followed, as well as how these lessons were incorporated into subsequent plans/programs.

State level officials of GSDMA were very clear that there has to be double-loop learning for success and they believe it is happening. One of them mentioned:

> The system keeps evolving. Still it continues. We keep learning, we keep upgrading. Our plans are being revised, equipment is added, more training is given to more people.

The respondents from municipality and local agencies were not so optimistic:

> There is no internalization of lessons in the municipality. We just keep doing the mandatory routine activities. (Municipality)

In fact, a political leader (municipal councilor) criticized the disaster management programs overall as ineffective and DEOC activities as lacking substance:

> All this is on paper only, not on the ground. They make nice reports. Computer has made preparation of reports, plans, and statistics very easy and colorful. (Councilor)

The surveys of government staff, however, showed that they generally agree that double-loop learning is happening, with high agreement in Bhuj (77 %) and low in Mandvi (50 %). Similar agreement was revealed for incorporation of lessons into plans.

All this shows that though there are good intentions to achieve double-loop learning, it is not happening practically, at the local levels.

Innovation

The interviews reveal that there is little innovation in government agencies. One of the municipal officers said:

> There is no innovation from staff as some are old and ready to retire; others are temporary workers.

However, it seems that NGOs and private businesses are trying to be more innovative in their practices.

> Yes, NGOs carried out effective awareness/education programs. (NGO)

> Cement companies send out mobile labs to test construction using their cement. This gives an extra validation of building safety. (Bhachau municipality)

In the survey, about 76 % of the government respondents agreed that there are innovative practices. However, respondents referred to the new institutional setup for disaster response as innovative practice, but did not refer to innovation as new local practices in response to the threat of future disasters. Contrary to the positive responses from government staff in Bhuj and Bhachau, Mandvi had only partial agreement with innovation occurring.

Overall, learning is low. The limited levels of flexibility/openness, information sharing, reach of education programs, and double-loop learning as well as practically no innovation render learning impossible.

6.5.5.2 Empowerment

Empowerment is the process of enhancing individual or group capacity to make choices and transform those choices into desired actions and outcomes (Gibson and Woolcock 2005). A group is empowered when disparate members of the group (i) know their rights and choices, (ii) express these in democratic means, and (iii) bring about the desired results. These are discussed below based on the interviews and surveys.

Know Their Rights and Choices

Awareness of rights seems to be low in the study area. The reasons vary from low literacy to the top-down approach of the government, which leaves no room for participation. A representative of an interest group suggested that the people get frustrated with government actions:

> Government actions take long time and people cannot wait that long. They then bypass government and take risks.

From the interviews, it became clear that one major issue with awareness is related to the external help received during disasters. Most communities received financial and material help from outside groups and the government. However, this made people assume that they will also receive such help in future. Therefore, they are not paying much attention to what their responsibilities are.

> People get outside help quickly. Therefore, no one worries about disasters or its aftermath. These days, they build better houses to save lives. They don't think about anything else. (ADA representative)

During the survey of residents of Bhuj, an overwhelming majority (90 %) disagreed that different choices for DRM plans are available to them and there were very minor differences in results based on gender, location, and income groups. Also, there were minor differences based on location of house. A majority (66 %) of community respondents disagreed that they are aware of choices and rights. While 71 % of women respondents disagreed that there are choices, only 46 % of male respondents disagreed. In addition, there were no differences based on location, given that there was general disagreement with the statement, there were some differences among different income groups. In particular, a majority of high-income group respondents agreed (67 %) that they are aware of their rights and choices, while others disagreed.

The results show that community members, in general, are not aware of their rights/choices; moreover, they do not feel that this issue is addressed properly by government. This could also be a problem associated with the implementation of programs.

Express Opinions in Democratic Manner

The interviews suggested that, in many cases, the rich and powerful caste groups decide the agenda for government actions. Since many neighborhoods are based on caste and religious groupings, within neighborhoods, the rich and influential elite assert their opinion to the whole group. This is evident in the participation level as well. Women and minorities have fewer opportunities for participation, although NGOs are working hard to bring them into the mainstream. At the city level, these strong groups tend to wield power over minorities and weak sections. Thus, development efforts tend to favor certain groups and locations of the city.

This finding is supported by the survey of residents as well. About 88 % of the community respondents disagreed that the process is democratic. Similarly, 90 % disagreed that the participation is high. Both results show that the DRM process in its current form is not democratic. This leaves the community with inequalities as the rich and powerful can divert the process to their benefit.

Bring About Desired Change

A majority (82 %) of community respondents surveyed disagreed that many DRM programs are implemented due to community pressure. Women respondents, on average, had high levels of disagreement with the statement. It is interesting to note that while a majority of respondents from all locations disagreed with the statement, slum respondents were divided in their opinion (50 % agreed and 50 % disagreed). This trend is consistent with differences between the income groups, with the lower-income groups having a more divided opinion than other groups. This could be understood from the fact that there are many NGOs carrying out programs that could be understood as related to DRM. For example, in one of the slum areas surveyed, an NGO had implemented a rain water harvesting system and helped the residents design a water distribution system. The NGO and residents also secured help from the local municipal councilor and other local leaders. Though this is a misconception, they might consider these activities as a government-led DRM initiative. However, this could not be considered a norm, rather it was an exception. A representative of one citizen group opined:

> There are no proactive view/actions from people. They are ready to complain but when it comes to taking action, they are not as enthusiastic. They expect government to do everything for them. On the other hand, whatever government is trying to do is not reaching the people.

6.5.5.3 Putting the Picture Together on Sustainability

While citizens, NGOs, and CBOs said that they are not aware of government actions and plans for disaster risk management, government respondents stressed that the community has not learned from the 2001 earthquake and continues to take no precautions against future disasters. A majority of interview respondents stressed that people have forgotten the 2001 earthquake and that they have not really learned much. Interviews also point to the dampening of capacity building momentum. An interview respondent working in a local NGO commented:

> Government was intensive for the first 3–4 years, not after that. Not many awareness programs happened after that. What is ADA doing? A local consultancy firm made a good City Development Plan. But ADA never implemented it or revised it properly.

Another NGO respondent commented:

> They [government] make colorful reports but nothing happens on ground. The stress was on rehabilitation and that was also in the first two years. After that, nothing much is happening. Even on rehabilitation, the third and last installment of aid is not yet distributed. Most of the time is wasted on meetings.

The establishment of the DEOC is the biggest sustaining factor since its core job is to coordinate disaster response and increase awareness. The GSDMA's demonstrated success and dedicated government budget also indicates sustainability. However, more involvement at the local level and higher interest from citizens is needed to sustain and enhance the momentum. The findings are summarized in Table 6.21 below.

6.5.5.4 Partial Capability Trap?

In the PeDJoLA model, sustainability is the effectiveness and continuity of local capacity even when there are changes in the personnel (agents). A capability trap occurs when there is an inability to achieve performance or desired collective goals

Table 6.21 Summary of learning and empowerment

Sustainability indicators	Government survey	Community survey	Interviews/overall
Learning			
Flexibility and openness to change	Yes	No	Partial, primarily in government
Information sharing and flow	Partially yes	No	Partial primarily in government
Double-loop learning	Yes		Partially yes
Innovation	Yes		Partially yes
Empowerment			
Know their rights and choices		No	No
Express in democratic manner		No	No
Bring about desired change		No	No

over time, even after implementing conscious capacity development efforts. Findings from this research suggest that capacity development efforts are only partially sustainable in the study towns, pointing to a partial "capability trap."

Based on the indicators developed for this field study, learning has taken place partially, primarily within the government system. Profound policy and organizational changes occurred after the 2001 earthquake, which have given birth to a whole new institutional structure on disaster management. However, the system is still very top-down, without much autonomy and flexibility at the local level. Learning has not been passed down to the communities, which are still not empowered to take the disaster risk management agenda further. Communities are still not receiving disaster risk management related information or are not using it. Moreover, they are neither aware of their rights and choices nor are they empowered to bring desired changes in terms of new projects or programs.

A capability trap leads to a situation where vulnerability to disasters remains high. Capacity development efforts, since they are not sustainable and entirely effective, provide a false sense of security in the study towns. In such as a situation, neither the government nor the community is doing enough to reduce disaster risk and prepare for disasters, and so communities remain at a high risk of disaster impacts.

In fact, after the GEERP, the study towns have attracted a lot of economic investment and have experienced increased population growth. While Bhuj's population decreased marginally before the 2001 earthquake (population decreased from 102,176 in 1991 to 98,528 in 2001), the population increased by 49 % between 2001 and 2011 (from 98,528 to 147,123) according to Census of India (2012). Thus, more people and assets are now at risk from earthquakes. Disaster risk still remains very high and preparedness low, while capacity development efforts are losing momentum with partial learning and no community empowerment—a situation of partial capability trap. The reasons for such a situation are discussed in the section below.

6.5.6 *Role of Agents and Factors Inhibiting Capacity Development*

What factors inhibit capacity development or lead to a capability trap? Primarily relying on the formal and informal interviews with change agents, this section describes the factors that inhibit capacity development or are leading to a capability trap in the selected towns. In the conceptual model, capability traps are related to (i) weaknesses related to organizational factors such as reliance on the form rather than on function, a closed system that does not provide flexibility and adaptability, and pressure to perform more than the existing capacity, thus leading to collapse; (ii) incentives and imperatives of change such as elite capture and rent seeking activities; and (iii) community factors, related to its empowerment and learning. Based on the conceptual model, the findings under this research question are organized under four major headings: (i) organizational factors, (ii) agents' incentives and imperatives, (iii) community factors, and (iv) particularly with respect to earthquake risk, continued structural vulnerability.

6.5.6.1 Organizational Factors

The GEERP was a very top-down project, leading to little connection with the local governments and communities. Formation of a new institutional structure parallel to the existing government machinery, although effective for emergency relief and reconstruction, is not very effective for disaster prevention. Existing local governments such as municipalities are directly answerable to the community for disaster preparedness, but are not charged with risk reduction in the new set up. Coordination is also a big challenge in the new set up since the new institutional structure has not yet been firmly established in the local governments. Not surprisingly, disaster preparedness is not a big priority for local government and the community. Coordination between different agencies also remains a big challenge. A respondent working in a local NGO commented:

> I don't see any coordinated activities among government agencies. There was better coordination right after the earthquake. It is no more there.

A respondent running a local private business who is also a member of an international philanthropic group said:

> The government is not well prepared. Even during minor floods, there is no real preparation or management. DEOC and city level agencies just conduct meetings and write reports but there is no follow up. About 4–5 years back, the people in the citizen committee were called for training. Nothing can be done in one meeting in 3–4 years.

The same respondent continued:

> We need more decentralization of disaster management and prevention activities.

A respondent working for a local NGO said:

> Not much coordination is happening now. There was coordination right after the earthquake. Some government leaders worked with NGOs at that time but nothing now.

6.5.6.2 Agents' Incentives and Imperatives

Several interviews are conducted with change agents who played key role in bringing change under the GEERP. These include political leaders; government staff at state, district, and local levels; and leaders from the private sector and interest groups, religious- and caste-based groups, and NGOs. The interviews point to various imperatives in the system, which inhibit their involvement in capacity development activities.

Political leaders Interview respondents said that politicians are not really fixing any fundamental issues but are just focusing on getting votes. Others pointed to the corruption in the system which undermines all the effort. Local NGO respondent mentioned:

> Politicians are only interested in getting votes and not in seriously fixing anything.

The following comment, which comes directly from a local political leader, is very shocking. He was sincere when he made this comment, pointing to the political system rather than his own behavior.

> Political leaders and government staff love corruption…You do whatever research, PhD, or report writing; we politicians will not change.

Government staff While technical staff complained about a very rigid system that does not help them respond effectively, internalize learning, or bring local issues at the front of the development discourse, many others pointed to a general lack of technical staff, good salaries, and rewards to attract talented and enthusiastic new talent. A local government officer noted that higher-ranked IAS officers are not genuinely interested in changing the situation as they are from a different region or state and usually only appointed to their position for three to four years. Newly formed ADAs were initially planned to be dissolved, but later were continued with limited resources. Many government officers pointed to these reasons for weak performance in managing illegal construction. A respondent from a local NGO commented:

> There may be genuine interest at higher levels. But as a system, there is not much change. Salaries are too low to attract good staff.

The political head of one of the ADAs echoed this theme:

> Our budget is very small. We have to meet our expenses from our revenue. Review fees and permit fees are the only revenue we have. There is dearth of staff too. The Chief Executive Officer is not full time and has additional responsibilities. If full time, we could have taken decisions quickly and taken up more activities. I have met the Chief Minister and Urban Development Minister and submitted letters, but nothing happened.

NGO respondent pointed to the commitment of staff and politicians:

> If officers are genuine, they respond well. The challenge is to make the authorities accountable. Politicians restrict their responses based on their party agenda.

Private sector and interest/religious groups, NGOs These groups are very small and have not taken disaster risk reduction and preparedness as their priority agenda. In response to their communities' needs, they played a major role in post-disaster relief and reconstruction phases but are not active in risk reduction. While NGOs cite limited funding in this field, private groups think that it is government's responsibility and they have no role to play.

6.5.6.3 Community Factors

Almost all the interview respondents stressed that the community's attitude is a major factor that inhibits capacity development. They stressed that people are not keeping preparedness and prevention as their priority because they have short memories, have other burning issues to focus on, do not value their own life as much as they should, do not want to take responsibility for their own life, follow a reactive way of responding to disasters, and rely on external funding from donors or rich groups for rescue and relief. A respondent from a reputable CBO said:

> People have a short memory—they forget the past and go on with life.

A political leader in a study town said:

> People are worried about their daily livelihood; other things come later. We all leave disasters to the God.

One of the mayors surveyed said:

> In general, people are not really better prepared than before to respond. In some cases, where people are exposed to new ideas and experienced the worst damage, people are aware and are hopefully better prepared. There are many newcomers in the city and they have no idea and are not interested too.

He further added:

> People try to save money by compromising on standards. People do not realize the value of their own life. This should change. Many of the so-called 'poor' have money, but wouldn't spend on a better house and life. They spend all that on vices such as alcohol, smoking, and gambling. That is their attitude. They need social actions to take them out of this. NGOs can do a lot in this area.

Another respondent from a local NGO said:

> Memory is too short. Not many people remember the 2001 earthquake now. Awareness about not to build or encroach upon all the open spaces is not there.

Many pointed to the reactive attitude of community and blamed both government and communities for not having healthy interactions. A respondent from a local CBO commented:

> Communities cooperate during emergencies but after that it evaporates.

According to a CBO representative:

> People forget and encroach upon new and open areas; government feels threatened with community involvement.

The mayor of a municipality said:

> People should also do things on their own. They cannot expect government to do everything.

There are also problems with external help and overall economic improvement. An interview respondent from a local NGO said:

> Disaster can happen any time due to our location. Post-2001 earthquake, rapid industrialization happened that brought in money to the local people. Also this arid region received good rains in the past 10 years. People became richer. They have forgotten the earthquake and its aftereffects.

Interview respondents painted a very bleak picture of citizens as selfish, valuing economic interest over safety, not taking responsibility of their own life, and relying on outside help. However, the interviews also alluded to some deeper reasons for this attitude. While higher- and middle-income groups are better-off in terms of their location and structural safety, the lower-income population, which makes up 46 % of the total population in Bhuj, still live in temporary or unsafe shelters/locations and do not have sufficient resources, skills, expertise, and awareness to invest in a better house. They have a limited say in government actions and thus have a feeling of powerlessness and seclusion. A respondent from a local NGO said:

> Building construction is controlled by a nexus of builders, politicians, and government staff who are supposed to enforce the rules.

A low-income group respondent belonging to the informal sector said:

> I have no choice but to stay in a known unsafe location because it is related to my daily earning. I cannot afford to lose this place as I have no other way of earning money.

Not trusting the government for timely help also leads to a sense of powerlessness and higher reliance on external support groups and God. This is evident from the response of one resident during an informal discussion:

> If there is a disaster, the local community helps private businesses in restoring their livelihood. But there is no support from the government. During recent floods, my merchandise was destroyed. I have not received any compensation so far from government.

6.5.6.4 Continued Structural Vulnerability

Apart from existing unsafe buildings, many new buildings do not comply with codes. Many old buildings still standing in Bhuj and Mandvi are unsafe and pose a threat to the nearby residents and users. Interview respondents were mostly optimistic in Bhachau town, which has gone through a complete reconstruction process. In Bhuj, buildings that were not damaged in the 2001 earthquake have now grown weaker due to lack of proper renovation and maintenance. With weak governance and supervision capacities in the local government, limited budget and technical staff, and masons following old techniques, the structural vulnerability is likely to rise in the future. With the increasing population trend, disaster risk is certainly going to increase.

A government officer in Mandvi said:

> The impact from a future disaster could be similar to the 2001 earthquake since there is illegal construction that includes building taller than allowed, sites fully covered (to mitigate high land prices), and low standard materials/techniques (to save a few pennies). There is about 300–400 % increase in land prices due to new developments such as roads, drainage, etc.

A representative from a citizen group in Bhuj said:

> There are at least 30 known old and big buildings that can fall anytime and pose an urgent threat in the most crowded walled city area. Government has not yet done anything to remove them.

A practicing engineer in Bhuj said:

> About 25 to 30 % of new construction does not follow current norms. Construction quality is also questionable.

These findings are summarized in Table 6.22 below.

6.5.6.5 Capacity Inhibiting Factors

Parallel, Top-Down Organizational Structure that Is Not Connecting with the Communities

In a very short period of time after the 2001 earthquake, national and state government supported district and local government machinery to provide immediate relief and reconstruction. However, there are many factors contributing to a continued capability trap. First, at the organizational level, a top-down, parallel institutional structure helped for the immediate disaster response but has hampered mainstreaming disaster management in development planning. This has also led to challenges in coordination and supervision of activities and maintenance of assets. The new institutional structure has shown results and legitimacy. However, it is losing momentum due to its disconnectedness with local governments and citizens. Second, agents have little or no incentives to keep prevention on their radar screen. Political leaders have moved on to other agenda. Government staff are temporary or technically less skilled and need better salaries and rewards to keep their motivation high. NGOs are already stretched thin, working heavily on many social aspects and are without funding in the area of disaster prevention.

Unmotivated Citizens and Change Agents

Interview respondents identified citizens as the biggest capacity development inhibiting factor. People have lost interest in disaster prevention and preparedness either as they have forgotten the 2001 earthquake, received large sums of money after

Table 6.22 Summary of factors inhibiting capacity development

Concept	Keywords	Themes
Coordination	No change, small	Top-down approach, agency coordination 50–50, no early warning, no DRM funds, coordination is less, no enforcement, DEOC is very small and cannot take care of activities across district
Technical/agent incentives—skills and resources	Deficiency, improved, no actions	No permanent staff in municipality. No special award for employee, staff and equipment deficiency, insufficient and overloaded staff, no awards for staff, no innovation, response side has improved but not proactive actions, no employee awards
Implementation	Not filtering, problem, no awards	State and local guidelines are not filtering to the community, implementation is a problem due to corruption, government staff not committed, no awards but some recognition, everybody is interested in DRM, but they just report
Community incentives	Attitude, change	People's attitude should change, leaving everything to God inhibits capacity, short memory, attitudes inhibit learning—pessimistic attitude of people and government staff; need continued interest and focus
Sustainability—learning	No internalization, sporadic, low	No internalization of lessons within municipality, system to transfer knowledge is missing, strong pressure groups and demands are needed, no long-term actions or awareness building, their efforts are sporadic, and coverage is very low
Sustainability—empowerment, leadership	Missing, left out	Momentum has slowed down, government is not involving NGOs, private sector, different groups; leadership and attitude are a big constraint, politicians are only interested in getting votes, slums are still left out

earthquake, do not value their life and property, or do not want to take responsibility for their own life. They have developed an attitude to rely on God and thus do not take any responsibility, saying "everything is fine and I don't need to make any changes." However, deeper analysis shows that this apathy is due to a general sense of powerlessness among those in the lower-income group, which constitutes more than 46 % of Bhuj's population.

High Structural Vulnerability

Finally, due to all the factors discussed above, vulnerability remains high. Illegal construction is continuing in the surveyed towns. Dangerous old buildings are still standing at various places, posing a safety threat. Some people are still living in temporary housing which is ill-fitted to face earthquakes. Disaster risks are not widely known and continue to increase with illegal construction. Effective early warning or alert systems do not exist. Masons who were earlier trained in safer building construction techniques are losing their skills.

Respondents have identified many factors that can improve the situation including decentralizing the disaster risk management program to the local level, creating larger awareness-building initiatives targeting citizens, giving a greater role to NGOs and CBOs, ensuring continuous deliberation and discussions on disaster prevention, and providing opportunities to improve technical capacity and working conditions of government staff, as well as creating greater linkages between the government and community activities.

The findings suggest important implications regarding the conceptual model. Capacity development and capability trap are along a continuum, similar to disaster resilience and vulnerability. Factors leading to a capability trap, as identified from the research, are (i) organizational structures that are very top-down with little flexibility and adaptability, (ii) no incentives for change agents, and (iii) communities that are not empowered or involved in learning. In the model, integration of top-down and bottom-up efforts contribute to effective and sustainable capacity development. The research findings show that this integration can come from involving community in capacity development efforts, empowering them, and focusing on joint learning. Continuous information exchange and deliberations will also integrate the government and community efforts.

6.6 Capability Trap at Local Level

Various interesting results emerged out of the field research analyzing local capacity development efforts in the three towns of Kutch in Gujarat State. The perceptions of various groups differ in terms of how they understand capacity, capacity deficit, and disaster risk. All groups believe that capacity was low before the 2001 earthquake, and they all agree that it has increased after the GEERP program. The state government, donors, and the community (NGOs, CBOs, businesses, and citizens) clearly linked the capacity deficit perception to the capacity development efforts just after the earthquake. However, a decade after the earthquake, most of the stakeholders believe that disaster risk is still high and capacity low but are either not taking any actions or taking partial actions to reduce perceived capacity deficits. To a certain extent, groups link their perception of capacity deficit to capacity development efforts only when sufficient resources are available and there are future benefits from capacity development efforts. Their tolerance for disaster risk is highly related to their perception of future benefit.

While the state and district administration think that their capacity has increased significantly, community groups and citizens believe that the government capacity has increased only to respond to a disaster, not necessarily with respect to preparedness. Government staff think that the capacity for disaster risk management has improved, except with respect to implementing building codes. While the state level government staff is more convinced that capacity has improved greatly, the local staff are relatively less convinced. This is particularly true with respect to the institutional (effective DRM plans), organizational (information sharing), and implementation (building codes) aspects of capacity. The community largely agrees that the government capacity has increased. However, while the NGOs, CBOs, and eminent citizens agree that it is good to have such a government structure as the GSDMA, they also believe that there is a need to keep the intensity of activities and build links with citizens. A majority of community respondents believe that their capacity is still low in terms of disaster risk management and prevention. While the abilities related to skills and resources, coordination, and leadership are low, cooperation and inclusion abilities are high. Interviews revealed that cooperation occurs primarily during disasters. All the respondents further point to the community's relaxed attitude for not paying attention to safe housing practices and for not putting enough pressure on the government to perform well in reducing disaster risk.

Based on the indicators developed for the Gujarat study, learning has taken place partially, primarily within the government system. However, learning has not passed down to the communities, and the communities are still not empowered to take the disaster risk management agenda further. Communities still do not receive disaster risk-related information nor are they aware of their rights and choices. They do not feel empowered to make desired changes in terms of new disaster risk management projects or programs. A majority of residents believe that citizens have not internalized the lessons from the 2001 earthquake and are not making sustained efforts for better preparedness or safety. Overall, the momentum and attention generated after the 2001 earthquake is dampening. The establishment of the DEOC is the biggest potential sustaining factor since its core job is to coordinate disaster response and increase awareness. The GSDMA's demonstrated success and dedicated government budget also indicates sustainability. However, more involvement at the local level and higher interest from citizens is needed to sustain and enhance the momentum.

Various factors point to a continued capability trap at the local level. At the organizational level, the top-down parallel institutional structure helped in immediate disaster response but has hampered mainstreaming it in development planning and has added to challenges in coordination, supervision, and maintenance. Second, agents have little or no incentives to maintain a focus on prevention. Political leaders have moved on to another agenda; government staff are either temporary or technically less skilled and need better salaries and rewards to keep their motivation high; and NGOs are small and without funding in the area of disaster prevention. Citizens have lost interest in disaster prevention and preparedness either due to powerlessness and seclusion from the whole capacity building process, or they are not aware of how their own actions are putting them at risk.

References

Argyris, C., & Schön, D. (1996). *Organizational learning II: Theory, method, and practice*. Reading: Addison-Wesley.

Asian Development Bank. (2008). *India: Gujarat earthquake rehabilitation and reconstruction implementation completion report*. Manila: Asian Development Bank.

BHADA. (2011). *Presentation on development of Bhuj after earthquake*. Unpublished presentation. Bhuj, Gujarat: Bhuj Area Development Authority.

Cardona, O. D. (2003). The need for rethinking the concepts of vulnerability and risk from a holistic perspective: A necessary review and criticism for effective risk management. In G. Frerks, G. Bankoff, & D. Hilhorst (Eds.), *Mapping vulnerability: Disasters, development, and people*. London: Earthscan.

Census of India. (2011). *Bhuj city census 2011 data*. http://www.census2011.co.in/census/city/317-bhuj.html. Accessed Dec 2012.

CEPT University. (2009). *Social impact assessment for the post-earthquake program*. Gandhinagar: GSDMA.

CEPT University. (2012). *Performance assessment system*. Available at www.pas.org.in. Accessed on 10th Dec 2012.

Drabek, T. E. (2005). Sociology, disasters and emergency management: History, contributions, and future agenda. In D. A. McEntire (Ed.), *Disciplines, disasters and emergency management: The convergence and divergence of concepts, issues and trends in the research literature*. Emmitsburg: Federal Emergency Management Agency.

Erramilli, B. P. (2009). *Disaster management in India: Analysis of factors impacting capacity building*. Unpublished PhD Dissertation, Georgia State University, UMI Number: 3406051.

Gazette of India. (2005). *The disaster management act*. New Delhi: Ministry of Law and Justice.

Gibson, C., & Woolcock, M. (2005). *Empowerment and local level conflict mediation in Indonesia: A comparative analysis of concepts, measures, and project efficacy*. Policy Research Working Paper No. 3713. Washington, DC: The World Bank.

Government of Gujarat. (2001). *A preliminary report on the earthquake damage in Gujarat*. Gujarat: Government of Gujarat.

Government of India. (2004). *Annual report 2003–2004*. New Delhi: Ministry of Home Affairs.

Government of India. (2005). *Report of the Administrative Reforms Commission*. New Delhi: Ministry of Personnel, Public Grievances, and Pensions. http://arc.gov.in/index.htm. Accessed on 10 Aug 2012.

Government of India. (2007). *The Constitution of India*. New Delhi: Ministry of Law and Justice.

Grindle, M. (1996). *Challenging the state: Crisis and innovation in Latin America and Africa*. Cambridge, MA: Cambridge University Press.

GSDMA. (2012). *Hazard profiles*. Available at http://www.gsdma.org/hazard.jpg. Accessed on 10th Dec 2012.

Jigyasu, R. (2002a). *Reducing disaster vulnerability through local knowledge and capacity: The case of earthquake-prone rural communities in India and Nepal*. Unpublished doctoral dissertation, Trondheim: Norwegian University of Science and Technology.

Jigyasu, R. (2002b). From Marathwada to Gujarat. Emerging challenges in post-earthquake rehabilitation for sustainable eco-development in South Asia. *Proceedings of the conference Improving Post-disaster Reconstruction in Developing Countries*. University of Montreal, Montreal, May 23rd–25th, 2002. Available at www.grif.umontreal.ca/pages/i-rec%20papers/rohit.PDF. Accessed on 16th Sept 2012.

Katiyar, S. K., & Khandelwal, R. (2001). Relocation in Kutch: Bridging old fault lines. *Economic and Political Weekly, 36*(26), 2319–2321.

Kenny, C. (2009). *Why do people die in earthquakes? The costs, benefits and institutions of disaster risk reduction in developing countries*. Policy Research Working Paper 4823. Washington, DC: The World Bank.

KPMG. (2005). *Benefit monitoring consulting sample survey report*. Unpublished power point presentation, KPMG.

Mehta, L. (2001). The manufacture of popular perceptions of scarcity: Dams and water-related narratives in Gujarat, India. *World Development, 29*(12), 2025–2041.

Nakagawa, Y., & Shaw, R. (2004). Social capital: A missing link to disaster recovery. *International Journal of Mass Emergencies and Disasters, 22*(1), 5–34.

NDMA. (2007). *National disaster management guidelines: Management of earthquakes*. New Delhi: National Disaster Management Authority.

Pritchett, L., & Weijer, F. (2010). *Fragile states: Stuck in a capability trap?* Input paper for The World Development Report 2011. Washington, DC: The World Bank.

Raju, K. C. B. (1995). *Strategies to combat drought in Kutch*. Workshop on strengthening of community participation in disaster reduction and role of NGOs, January 13–15, New Delhi/Mandvi: Vivekananda Research and Training Institute.

Sanderson, D., & Sharma, A. (2008). Winners and losers from the 2001 Gujarat earthquake. *Environment and Urbanization, 20*(1), 177–186.

Sato, T., Hamada, M., & Hayashi, Y., et al. (2001). *A comprehensive survey of the 26 January 2001 earthquake (Mw 7.7) in the State of Gujarat, India*. Report by the Research Team supported by the Grant-in-Aid for Specially Promoted Research Provided by the Japanese Ministry of Education, Culture, Sports, Science and Technology in the fiscal year of 2000. Available at http://www.st.hirosaki-u.ac.jp/~tamao/Gujaratprint.html. Accessed on 13th Dec 2012.

Sharma, M. B. (2009). Disaster as an opportunity for development of social impact assessment of Gujarat Earthquake Emergency Reconstruction Project (GEERP). *National Seminar on Housing and Disaster Mitigation Efforts in India: Case Studies*. Available at http://bhagirathi.iitr.ac.in/dspace/bitstream/123456789/936/1/2%20Madhu%20Bharti.pdf. Accessed on 10th Nov 2012.

Sheth, P. N. (2000). Gujarat. In G. Mathew (Ed.), *Status of Panchayati Raj in the States and Union Territories of India 2000*. New Delhi: Concept Publishing Company for Institute of Social Sciences.

Simpson, E. (2008). *The rise and fall of collective public action in the aftermath of the Gujarat earthquake of 2001*. NGPA Working Paper Series. London: London School of Economics and Political Science.

Thiruppugazh, V. (2007). *Post-disaster reconstruction and the window of opportunity: A review of select concepts, models and research studies*. JTCDM Working Paper Series No. 3. Mumbai: Jamsetji Tata Centre for Disaster Management, Tata Institute of Social Sciences.

UN-HABITAT. (2009). *Land and natural disasters: Guidance for practitioners*. Nairobi: United Nations Human Settlements Programme (UN-HABITAT).

UNISDR. (2004). *Living with risk: A global review of disaster reduction initiatives*. Geneva: United Nations International Strategy for Disaster Reduction.

UNISDR. (2009). *Global assessment report on disaster risk reduction*. Geneva: United Nations.

UNISDR. (2011). *Global assessment report on disaster risk reduction*. Geneva: United Nations.

World Bank. (2005). *Capacity building in Africa: An OED evaluation of World Bank support*. Washington, DC: The World Bank.

World Bank. (2009). *Implementation completion and results report on Gujarat Emergency Earthquake Reconstruction Project*. Washington, DC: The World Bank.

World Bank. (2010). *Natural hazards, unnatural disasters: The economics of effective prevention*. Washington, DC: The World Bank.

Yagnik, A., & Sud, N. (2005). *Hindutva and beyond: The political topography of Gujarat*. London: LSE Crisis States Research Centre.

Chapter 7
Breaking the Capability Trap

Abstract What can be done to break the capability trap? Two radically different approaches will be discussed in this chapter. First, an internal and community centric approach driven by innovative ways of financing and no or little reliance on donor funding. Second, based on changes in structural design, monitoring, and implementation of donor or top-down aid. In both the approaches, communities need to be at the center of capacity development efforts. Many country governments are now realizing this and are focusing their efforts on involving communities earlier on in their disaster risk management programs. At the same time, local government agencies such as municipalities and development authorities need to be empowered to lead the disaster risk management agenda. Higher level governments and/or donors can facilitate this change by providing technical, financial, and decision-making power to lower levels of governments. The critical roles of local government officers and community leaders—the change agents—need to be supported by rewarding them for their innovative practices and dedication. The chapter starts by discussing the implications of the Gujarat study on literature and practice. A review of international cases is presented to understand how disaster risk management system has evolved across countries. This is followed by a set of recommendations for practitioners and researchers.

Keywords Behavioral change • Empowerment • Intervention design • Community-centered risk mitigation • Perception-based learning

What can be done to break the capability trap? Two radically different approaches will be discussed in this chapter: first, an internal and community-centric approach driven by innovative ways of financing and no or little reliance on donor funding and, second, based on changes in structural design, monitoring, and implementation of donor or top-down aid. In both the approaches, communities need to be at the center of capacity development efforts. Many country governments are now realizing this and are focusing their efforts on involving communities earlier on in their disaster risk management programs. At the same time, local government agencies such as municipalities and development authorities need to be empowered to lead the disaster risk management agenda. Higher-level governments and/or donors can

A. Tiwari, *The Capacity Crisis in Disaster Risk Management*,
Environmental Hazards, DOI 10.1007/978-3-319-09405-2_7

facilitate this change by providing technical, financial, and decision-making power to lower levels of governments. The critical roles of local government officers and community leaders—the change agents—need to be supported by rewarding them for their innovative practices and dedication. The chapter starts by discussing the implications of the Gujarat study on literature and practice. A review of international cases is presented to understand how disaster risk management system has evolved across countries. This is followed by a set of recommendations for practitioners and researchers.

7.1 Capacity Development After the 2001 Earthquake in Gujarat: An Overview of Findings from Gujarat Towns

The PeDJoLA model of local capacity development for preventing and preparing for disasters was used to understand the disaster risk management capacity developed under the Gujarat Emergency Earthquake Reconstruction Project, primarily in three towns of Kutch in Gujarat state, India. The study, primarily through interviews and surveys in three towns of Gujarat, was an attempt to understand how a typical donor and federal/state government capacity development program evolves over time. At the same time, it provides an opportunity to understand factors for effective and sustainable capacity.

The conceptual model identified two major groups of stakeholders in the capacity development process. First, those with resources and responsibility for developing and implementing a capacity development intervention are represented by the donors and government, primarily following a top-down approach. At the receiving level are communities, who are not just passive recipients of interventions but are dynamic social groups networked with each other and engaged in bottom-up interventions. The capacity building process takes place at a local level within an environment or disaster risk landscape. This environment or landscape is defined by the physical location, climate, and geophysical characteristics as well as socioeconomic characteristics. These socioeconomic characteristics include community structure, political dynamics, and administrative/governance issues. This landscape plays important roles in shaping implementation process and its results. The process starts with a perception of capacity deficit or disaster risk, which then leads to formulating top-down and/or bottom-up interventions. Change agents such as government officers who actually design and implement programs, political leaders, and community leaders play critical catalytic roles in leading transformational change.

The entire capacity building process works along a continuum. At one extreme, the combined efforts can lead to sustainable capacity building and disaster risk reduction. On the other extreme, the efforts can lead to a capability trap or continued vulnerability to disasters. Sustainability and effectiveness of capacity development efforts are based on how integrated top-down and bottom-up interventions are, how well the community is involved in learning and empowerment, and how much the interventions contribute to improving the environment itself. A capability trap can result from a program that does not provide flexibility, adaptability, and community empowerment; organizational factors such as reliance on the form rather than on function; and elite capture and rent-seeking activities of change agents that lead to no public value creation or transformational learning to drive change. A feedback loop links the outcome of capacity development efforts to perception, which is then linked to interventions. The entire process is dynamic, with multiple feedback loops linking various stakeholder groups from perception to actions and then results.

The field research, based on the conceptual model, provided an insight into the capacity building process and capability trap situation. All interviewed groups agreed that capacity was low before the 2001 earthquake and increased after the GEERP program. The state government, the donors, and the community (NGOs, Community-based organizations (CBOs), businesses, and citizens) clearly linked the capacity deficit perception to the capacity development efforts just after the earthquake. However, a decade after the earthquake, most of the stakeholders believe that the disaster risk is still high and capacity low but that there are either few or very few actions to reduce perceived capacity deficits. While the state and district administrations think that their capacity has increased significantly, community groups and citizens believe that the government capacity has increased only to respond to a disaster, not necessarily with regard to preparedness and prevention. A majority of community respondents believe that their capacity is still low with respect to disaster risk management and prevention. All the respondents further point to the community's laid-back attitude, not paying attention to safe housing practices and not putting enough pressure on the government to perform well in reducing disaster risk.

Based on the indicators developed for the Gujarat study, learning has taken place partially, primarily within the government system. Learning has not, however, passed down to the communities, which are still not empowered to take the disaster risk management agenda further. Various factors point to a continued capability trap at the local level. At the organizational level, a top-down, parallel institutional structure helped in immediate disaster response but has hampered mainstreaming in development planning and added to challenges in coordination, supervision, and maintenance. Second, agents have little or no incentives to stay focused on prevention activities. Political leaders have moved on to other agenda; government staff are either temporarily or technically less skilled and need better salaries and rewards to keep their motivation high; and NGOs are small and without funding in the area of disaster prevention. Citizens have lost interest in disaster prevention and preparedness either due to feelings of powerlessness and seclusion from the whole capacity building process or because they are not aware of how their own actions are putting them at risk.

Overall, the study does not find major differences in the responses of government officers from the three towns regarding government capacity for disaster risk management. Interviews show that the government officers in Mandvi, which was the town that was least affected by the 2001 earthquake (among the three study towns) and which received the least support under the GEERP, disagreed on the existence of implementation, technical, and political capacity for managing disasters. The interview respondents from government officers of Bhuj and Bhachau, which were highly affected by the disaster and had massive support components under GEERP, disagreed that there is a sufficient capacity for implementation and institutional capacities (implementation of building codes and existence of effective disaster risk management plan). The study also shows that perceptions of various groups differ. Within the government responders, the local government staff differed from the state and district officers. Within the community, perceptions of lower-income group citizens differed from middle- and high-income groups, while opinions of men and women responders only differed slightly.

7.2 Reflecting on the PeDJoLA Model

The field research conducted in Gujarat has important implications for the conceptual framework presented in Chap. 5. Overall, the conceptual model provides a good starting point for evaluating capacity for managing disasters. The framework links the perception of risk and capacity with intervention design, change agents, environment, system, and results. The research also shows some challenges in applying this framework including how to evaluate community capacity, how to assess integration of community and government capacities, and what is meant by capacity development and capability trap in the real world. These points are further discussed below through the main components of the model.

7.2.1 Disaster Risk, Current Capacity, and Capacity Deficit Perception

The field research shows that different stakeholders have different perceptions of disaster risk and capacity. People's tolerance for risk, in particular, dealing with disaster risk, current capacity, and capacity deficit perception, is based on their perceived benefits. The research supports both these points. However, the second point is more relevant in a "normal situation," many years after the disaster is over. In the study towns, just after the earthquake, almost every group had similar perception of needs, which compelled all the groups to take some actions. However, over time, the perception has changed with different groups perceiving disaster risk and capacity in different ways. A common understanding or perception helps in developing integrated and comprehensive interventions.

At the top-down level, perceptions of donors and governments related to disaster risk are largely driven by past studies, government reports, and media. At the community level, the findings suggest that perceptions of different community groups differ and citizens who are aware of disaster risk have different levels of acceptable risk that affect their preparedness levels. An individual's perception of future risks and benefits lead to his/her disaster risk perception.

Perceptions of donors and governments regarding current capacity or capacity deficit were largely based on performance indicators set up for capacity building programs. If the numerical targets are achieved, capacity is assumed to be in place. This has been mentioned in the literature also. However, at the donor level, the sustainability of project outcomes was questioned. At the community level, the findings suggest that for the most parts, citizens, NGOs, CBOs, and the private sector had a similar understanding of current capacity and capacity deficit areas.

7.2.2 Intervention Design

This part of the model links disasters risk, current capacity, and capacity deficit perceptions with capacity development actions. Findings from the field research suggest that the link between capacity deficit perception and actions is not straightforward and varies with time and stakeholder groups. While just after the disaster, the capacity deficit perception is more likely to be translated into capacity development efforts, as time passes (3–4 years in Gujarat), such a link becomes invisible and capacity development momentum is lost. Other studies, such as those by Jigyasu (2002) and Jha et al. (2012), have also mentioned about disaster's "half-life." The current perception of the majority of respondents in the study towns is that there is no urgency to take any actions since needed actions do not necessarily indicate any present or future economic or social benefits. The majority of respondents perceive that the next earthquake is most likely to happen far into the future (50–70 years from now). If there is a disaster, the community expects to receive outside support to rebuild as happened after the 2001 earthquake.

Higher-level government officials and donors are more likely to take actions based on capacity deficit perceptions immediately after a disaster, not so much after some time has passed. In the case of Gujarat, the Gujarat State Disaster Management Authority (GSDMA) structure still provides sustainability to the disaster risk management program. At the community level, just after the disaster, people, NGOs, and CBOs contributed to the best of their ability. A decade after the disaster, they are not taking any measures to reduce the perceived capacity deficit. The reason may lie in the GEERP design not supporting a better or more enabling environment, community learning, and empowerment. Ensuring a common perception to disaster risk and capacity (through advocacy and awareness programs), ensuring joint learning, involving local governments and communities in decision-making, and providing incentives for investing in disaster risk mitigation and preparedness help in ensuring transformational change, which improves the environment and reduces disaster risk.

7.2.3 Capacity Development and Implementation Process

The field research in the three towns of Kutch shows that the capacity development process was largely top-down in Gujarat. Brinkerhoff's model (2008), relying on resources, skills and knowledge, organization, politics and power, and incentives, is partially supported by the findings of this research. In Gujarat, the government and donors had resources, knowledge, and power to develop and implement the GEERP. Grindle's (2007) model also helps to explain the details of how capacity building takes place on the ground from agenda setting to design, adoption, implementation, and sustainability. At the bottom-up level, however, no other theory fully explains the situation. The community structure consists of many groups with differing perceptions of risk and capacity and different levels of acceptable risk. All the groups, however, look to the government to take a lead in all the aspects of disaster risk management. The GEERP program only partially involved community at the grassroots level and thus has achieved limited traction with communities.

The field research assessed government and community capacity based on various indicators and found that although government capacity has increased significantly, community capacity has only partially increased. Better community capacity indicators are needed to better capture networking and social capital analysis necessary for disaster risk management. Interviews with change agents provided better insight into the process. Perhaps there is no better way to judge top-down and bottom-up interventions than in-depth interviews with change agents.

7.2.4 Role of Agents

The fieldwork confirmed that agents indeed play an important role in realizing the capacity development efforts. They indeed work in a landscape of incentives and imperatives. However, in a top-down framework, local agents such as community leaders, CBOs, and politicians have a limited say and involvement regarding the capacity building efforts. The research findings did show that they work within an environment with multiple ongoing tensions—imperatives and incentives that characterize the space and that either reward or inhibit innovation (Pritchett and Weijer 2010). While the research found very few incentives for government staff to bring change, since there is limited flexibility and empowerment of local government staff, the study also found that leadership played a key role in initially setting up GSDMA and implementing reconstruction plans.

7.2.5 Role of Environment and System

In the model, both the government and the community capacity exist within a dynamic system. The system exists in balance with its environment. The environment consists of economic, political, administrative, and sociocultural characteristics (Brinkerhoff 2004). In this research, disaster risk and the environment were considered similar. In many ways, existing vulnerability and exposure to hazards shaped the outcome of results. For example, community vulnerabilities such as structural vulnerability of houses, land use, and infrastructure still continue, even after undertaking massive reconstruction programs. Community structure played an important role in getting resources for relief and rehabilitation. Politics and administrative structures also played an important role in how interventions were designed and implemented.

Systems, on the other hand, need flexibility to adapt to changing situations and involve relationships between the parts and the whole, between the system and the environment, and within different parts (Wildavsky 1988). In this case, the system is a disaster risk management system set up under GSDMA, which includes a larger system working at the state level, and has links to the national, district, taluka, and local levels. The parts are the constituent local areas as well as administrative units, which work together to achieve a common goal—disaster risk management. Information flow within the system indeed plays an important role to link different government officers and the community. The research shows that the disaster risk management system is not very flexible and its linkages at the local level are relatively weak compared with the linkages at the district level. Enhanced focus on disaster risk reduction is needed, which primarily requires reducing structural and community vulnerability at the local level.

The formation of a new institution affected the environment itself, improving the administrative structure to manage disasters. Although its linkages at the local level are questionable, overall, the GSDMA improved the environment and the system, enhancing the environment and reducing disaster risk. Other forms of capacity development efforts such as reconstruction and training also contributed positively to the environment and reduced disaster risk.

7.2.6 Current Capacity, Capability Trap, Risk, and Vulnerability

In the conceptual framework, capacity development for disaster prevention and response is a locally driven process, involving transformational learning by local organizations, community, leaders, and other agents, which leads to disaster resilience (World Bank 2011). Many interview respondents agreed with this definition during the fieldwork. However, this type of transformational learning is not happening at the local level in the study towns. Capability trap, on the other hand, is defined as an inability to achieve performance or desired collective goals for a long time, even after implementing conscious capacity development efforts (Pritchett and Weijer 2010). Such inability will be related to weaknesses along one or more dimensions of local government and community capacity. The GEERP has shown progress and results when measured through numerical indicators established by the donors. No desired collective goals were established for the GEERP that included communities and local governments. Evaluated through combined government and community capacity, the research showed that GEERP has resulted in a capability trap. The GEERP is based on an institutional structure with effective and very impressive capacity at the state level but that shows weakness at the local level.

Discussion from a risk and vulnerability point of view provides more insight into the capability trap situation. The primary goal of the GEERP was to manage reconstruction after the 2001 earthquake. Disaster risk reduction was a secondary goal, attached to the reconstruction program based on donor recommendations. However no real goals were fixed for disaster risk reduction. It is noteworthy that disaster committees, which were supposed to be established in all communities, merely exist on paper. The most important contribution of the GEERP toward risk reduction was to follow a "build back better" approach, which helped in developing risk-sensitive land-use planning and building construction. After the reconstruction program was over, the GSDMA assumed a wider role for disaster risk reduction with a focus on undertaking disaster risk assessments, setting up an early warning and response system, undertaking disaster risk awareness programs, and training people on various disaster risk management topics.

Disaster risk has accumulated in the three study towns. Population and building stock and other physical assets have increased considerably in the study towns such as Bhuj. The fieldwork shows that while the houses reconstructed as part of GEERP are earthquake resistant, many new and old houses (which were not affected by the 2001 earthquake) are still vulnerable. New citizens who were not exposed to the 2001 earthquake have little or very little motivation to invest in risk reduction. In part, this is due to the lax supervision of building codes by the area development authorities. The ADAs were set up to help with urban development planning and implementation after the 2001 earthquake. However, they are now in need of financial and technical resources to keep pace with the rapid development. Their functions are separate from those of the municipalities, which are politicized and run by elected representatives. Initially the state government wanted to dissolve these ADAs. However, they continued but without provision for adequate resources.

Many interview responders predicted fewer fatalities but potentially higher damages in the next earthquake. This situation points to a capability trap, where there are institutions in place but without adequate capacity to actually manage the situation. However, there is great uncertainty about earthquake events—nobody can predict if there will be an earthquake or what will be its intensity. The study towns are more likely to be affected by droughts and floods due to climate change. Are they ready for those events? Timely warning for these events can be provided. However, contingency plans still need more coordination and dissemination.

7.3 What Can Literature Tell About the Findings from the Gujarat Towns?

The study of Gujarat towns has important implications for both the capacity building and disaster risk management literature. As discussed in Chap. 3, the concept of capacity development forms a part of the growing list of esteemed concepts used by development practitioners to solve development challenges. The literature in this field is full of contentious debates and dilemmas, with new research pointing to a concept that is the opposite of what it is intended for—capability traps. The case study analysis of the capacity building programs undertaken by international donors in Gujarat after the devastating earthquake of 2001 confirms a number of themes identified in the literature review. First, similar to the research conducted so far on the performance of capacity development programs (World Bank 2005; Grindle 2007; UNESCO 2009), it paints a mixed picture. Second, as shown in the literature, the capacity development under the GEERP utilized typical donor-driven tools and approaches such as best practices and supply-driven and short time-frame projects focusing on fiscal disbursement and quick fixes. Clearly the GEERP design was primarily based on the initial assessment of the situation by the World Bank and ADB and incorporated lessons learned or best practices from the banks' operations in other countries as well as in India. The program itself was designed top down and in a rapid manner to respond to the disaster situation.

Particularly, the formulation of new institutions such as GSDMA is similar to the formation of other new institutions for capacity development, which are expected to perform "too much of too little too soon too often" (Pritchett and Weijer 2010, p. 37). As per Pritchett and Weijer, such institutions crush the system, leaving it in a "paralyzed" state where it mimics the form on the surface but remains weak underneath. In the case of Gujarat, the GSDMA has gathered a lot of credibility and has established itself firmly in the state's administration framework. However, at the local level, the top-down, parallel, and disjointed institutional structure of the GSDMA, though helpful in the immediate disaster response, has not able to achieve the mainstreaming of disaster risk management in development planning. The capacity of local government institutions remains limited since the new institutional structure of GSDMA is parallel to the established functions of municipalities. Thus, community empowerment as well as joint learning is not taking place. In many ways, the situation is very similar to the assumptions of Pritchett and Weijer, as well as of Andrews. To a certain extent, GSDMA shows "isomorphism, institutional dualism, and flailing states" (Andrews 2008b, p. 4). Isomorphism is the tendency of organizations to imitate characteristics that are generally considered effective while the organization has no real capacity (Andrews 2008b, p. 33).

Institutional dualism leads to government systems and processes running parallel to each other leading to a situation in which governance has no bearing on how public decisions are actually made and implemented (Andrews 2008b, p. 28). The concept of "flailing state" is similar to institutional dualism, in which government introduces a parallel system of new and best practices on top of a preexisting system, with no connection between the two, leading to general confusion and failure (Andrews 2008b, p. 4). The GSDMA acquired all the characteristics of an effective institution, which is not necessarily bad. In fact, establishing such an agency to coordinate donor responses and actively undertake rapid reconstruction was necessary in the post-disaster situation. The challenge is more related to its role in the post-reconstruction period. At the local level, the GSDMA cannot control disaster risk arising out of physical vulnerability because ADAs are in charge of issuing building permits and ensuring that buildings follow prescribed codes but are not responsible for disaster risk management. While Pritchett and Weijer have based their assumptions on fragile states, Gujarat's vibrant economy coupled with a strong culture of mutual help and a good public administration has played a big role in delivering results for the GSDMA.

In terms of the conceptual models and theories of capacity development, this research elaborates concepts put forth by Andrews (2008a) and Pritchett and Weijer (2010). However, when designing a new capacity development intervention, Grindle's (2007) conceptual framework, which relies on assessing context, contents, and processes of reforms, is a more useful framework. Based on the new conceptual model, Grindle's conceptual framework can be improved by adding an evaluation of local government and community context, as well as contents and processes of reforms to be able to develop a meaningful intervention at the local level. The application of the new conceptual model has helped in explaining factors related to actual capacity building, how institutions evolve, and how local development should be defined—a gap identified by many scholars (Easterly 2008; Pritchett et al. 2010). Some questions which can further refine the new model are as follows: How can community capacity be analyzed comprehensively? How should environmental and system factors (context as defined by Grindle) be defined? How can capacity or capability trap of various groups and subgroups that work in formal (government system) and informal (community) settings be better understood?

The field of disaster risk management is based on a new understanding that disasters are a product of natural and social forces. However, as discussed in Chap. 4, even with increased efforts and understanding associated with disaster response and mitigation, monetary losses from disasters are increasing globally. This research provides some justification for this trend. Although the government has invested heavily in disaster risk management, it is likely that, in the next earthquake, the three study towns will suffer higher damages than those seen during 2001. This is due to population growth, improved economy (which leads to higher assets), and increased investment in buildings and other physical infrastructure during the reconstruction. The research also provides some implications for the literature, particularly related to the concepts of disaster risk, vulnerability, roles and responsibilities of major actors, and to a comprehensive theory of risk and resilience.

The field research shows that the new concept of disaster risk is still not understood by local communities. While at the state levels, disaster risk maps are being prepared, communities are still not aware that their own vulnerability and actions play a big role in shaping their disaster risk. Many community responders believe that the disasters are an "act of God," and they have no control over it. Beck (2006) described this thought process during pre-modernity period when the prevalent attitude about disasters was that mankind has no or limited power to prevent or change anything. During that era, the focus was on relief and response. Before the 2001 earthquake, even the government had such a belief with little focus on risk reduction and a massive response program for earthquake relief. However, after 2001, there was double-loop learning for the government, which prompted investment in disaster risk reduction. Thus, new disaster risk maps are being prepared including seismic microzonation. To a certain extent, this thought is similar to those held in the industrial era (Beck 2006), when the government wanted to show that it is in charge and disaster risk could be measured and controlled. Clearly, the community has yet to reach that understanding. In reality, the society as a whole has moved toward radicalized risk (Beck 2006). More people and assets are now exposed to disasters in the study towns, new industrial areas have sprung up, and there is uncertainty about when, where, and at what intensity the disaster will strike. Risk assessment would help in modeling disaster impacts during different scenarios of disasters. However, the community demographics and built structures are changing at such a speed that risk assessments will find it hard to catch up with the increasing risks. Thus, the study towns are simultaneously living in all three eras as depicted in Fig. 7.1 below.

The research also provides insight into the concept of vulnerability, which is defined as a characteristic of persons or groups in terms of their capacity to anticipate, cope with, resist, and recover from the impact of a natural hazard (Wisner et al. 2004). The study has shown that perception and overall understanding regarding disaster play a key role in preparing and recovering from disasters. The field research shows that a majority of local stakeholders never anticipated an earthquake of such a big intensity as that of 2001. The interview respondents stated that such a risk seemed much lower before 2001. However, after the devastating earthquake, most

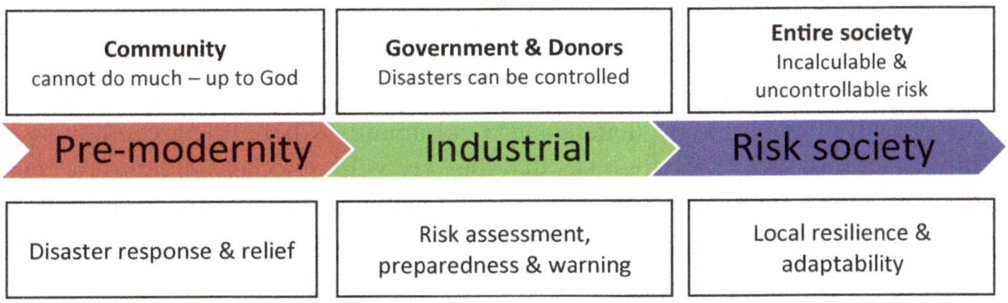

Fig. 7.1 Current status and future of beliefs and actions in study towns (*Left two boxes* show beliefs and current actions of the community and government in the study towns. The *rightmost boxes* show what the reality is and needed actions)

of the community knows that the towns are under seismic zone V and devastating earthquakes are possible. Still they are not taking required actions due to their belief that disasters are an act of God and other factors such as their lack of financial resources. Thus, the starting point of engagement with the communities should be to understand their belief system, which may not be necessarily easy to change. This should then drive the work on increasing their awareness about the current understanding of disasters. Even after understanding disaster risk, citizens may not take actions. A research conducted in the USA helps in explaining who prepares for disasters but does not necessarily tell why (Mileti and Fitzpatrick 1993). The reasons may be related to the current and future benefits of preparedness. This would justify why a vegetable seller in Bhuj would still like to sit under a crumbling building even after knowing that an earthquake can happen anytime—since his daily earnings are more important to him than the risk of losing his life.

One of the important areas of the disaster risk management literature is to understand the roles of different actors such as households, groups, organizations, and governments during different phases of a disaster. The research on disaster risk management thus far has focused on how different actors should behave and to some extent on how they are actually behaving. Research conducted in the USA shows that only some households prepare and others do not; post-disaster housing patterns reflect pre-disaster social ties, conflict, and socioeconomic profile; and many citizens volunteer only in the immediate aftermath of a disaster. Research on the roles of the private sector and media is still relatively limited. The fieldwork in the three study towns shows that citizens, NGOs, private sectors, and media are not fully aware about their roles and responsibilities during different phases of disaster risk management. The state and district governments have developed adequate capacities as needed for different phases. However, they exist only partially at the local government level. Table 7.1 below shows the normative roles of different actors in different phases of disaster risk management compared to those that are currently observed in the study towns. Statements written in black show normative roles during different phases, and those in red show those observed in the three study towns.

The field research has important contributions for the literature on disaster risk management in cities of developing countries, particularly in the following areas:

- *Urban land-use planning and risk reduction*: Land-use planning can reduce disaster risk (Sengezer and Koc 2005). However, to ensure risk-sensitive land-use planning, urban planning and effective risk assessment should start with community campaigns to understand a community's beliefs, perceptions, and acceptable levels of risk (the risks that they are willing to bear in return for current or future benefits). Anything planned without considering these will be futile since the community will not be willing to maintain or invest in risk reduction measures. At the same time, such a campaign will likely change a community's understanding of risk and its willingness to take proactive actions. Government incentives and supervision should follow such campaigns.

Table 7.1 Normative roles of different actors in DRM literature and situation in Gujarat

	Preparedness	Response	Recovery	Mitigation
Households, families	Have emergency kits; know what to do during emergencies [*No emergency kits, some awareness of what to do during emergency*]	Follow emergency plan [*Emergency plan is not known to all and is not discussed regularly*]	Draw on stocks of social credit; temporal migration [*Yes*]	Owning multiple assets; investments to protect and maintain assets; migration; property and life insurance [*Majority of low- and middle-income-group citizens have no mitigation measures*]
Community groups and nonprofit organizations	Emergency kits; knowing what to do during emergencies [*No emergency kit, some awareness of what to do during emergency*]	Informal warning [*For cyclone and floods only*]	Enhance social capital and help in community recovery [*Yes*]	Relocation to safe areas; community training programs; community-based information system; other infrastructure [*Some relocation occurred, community training program exists*]
Private sector	Emergency kits; knowing what to do during emergencies [*No emergency kit, some awareness of what to do during emergency*]	Actively participate in emergency plan; prepare a plan for their business [*None*]	Existing recovery plan [*No existing recovery plan*]	Private disaster insurance [*None*]
Media	Emergency kits; knowing what to do during emergencies [*No emergency kit, some awareness of what to do during emergency*]	Actively participate in emergency plan; prepare a plan for their business [*None*]	Existing recovery plan [*No existing recovery plan*]	Education programs; awareness [*None, except occasional coverage on disasters*]
Government National State Local	Formulating, exercising, and testing disaster contingency plans; training responders and the general public; public awareness campaigns about potential problems during disasters and what to do [*Existing at state and district levels, weak at local level (contingency plans are not discussed/disseminated)*]	Emergency shelter; search and rescue; damage assessment; coordination, communication, resource mobilization [*Existing at state and district levels, partially at local level (partial coordination and awareness at local level)*]	Prepare recovery and reconstruction plan; implement and monitor [*Existing at state and district levels, partially at local level*]	Disaster risk profiles; disaster-resilient infrastructure and buildings; adequate social and physical infrastructure [*Partially existing at state and district levels, (disaster profiles available, huge task of risk reduction at local level still needed)*]

- *Learning from past disasters*: A research conducted in Turkey and India (Orissa state) shows that government does not seem to learn from past disasters (Corbacioglu and Kapucu 2006; Winchester 2000). The fieldwork in Gujarat shows that it is, in part, because the focus has not been on joint learning with the communities. In Gujarat, the capacity development intervention in disaster risk management was undertaken after a devastating disaster. The focus was on reconstruction, with some elements of prevention and mitigation through a state agency. While state capacities are necessary to provide long-term risk reduction efforts, empowerment of local governments and communities is also necessary to ensure flexibility, adaptability, and effective change. After all, community resilience will ultimately determine impacts from the next disaster. The local government and community together need to understand and take steps for investing in prevention, mitigation, and preparedness. This understanding will only come if the communities are motivated, involved, and active in joint learning with the government—what are the risks, what can be done, which actions are successful and which are not, and how can they adapt to a new risk. The government's role will then be to provide information, facilitate dialogue between various community groups, ensure equity, and serve as custodians of lessons learned so that they can be incorporated in land-use planning and other tools of managing risks. Empowered communities, in return, are likely to keep pressure on the government to keep the risk reduction agenda active.
- *Reconstruction policy focusing on reducing structural risk (on buildings and infrastructure) increases overall disaster risk*: A research conducted in the Indian state of Orissa and in Sri Lanka regarding the impacts of post-disaster reconstruction policies suggests that the pressure to quickly rebuild and focus on reducing structural vulnerability may actually increase the long-term vulnerability of the affected community (Winchester 2000; Ingram et al. 2006). This is also true for the three study towns. Land-use planning and "build back better" reconstruction practices reduced structural risk significantly when the GEERP was active. However, currently the loss of earthquake-resistant construction skills by many masons, lax building control measures, and lack of awareness or motivation by the communities are putting the new and very old infrastructure at risk of damages from future disasters. At the same time, reconstruction has attracted more investments and an increase in the population, resulting in higher likely exposure to future disasters.

Finally, this research contributes to a comprehensive theory of risk and resilience (Cardona 2003) by stressing that communities have a big role to play in disaster risk management. Joint learning by the government and communities can be encouraged by understanding communities' beliefs, perceptions, and their acceptable level of risk; increasing their awareness; involving the local governments to lead disaster risk management agenda; and providing flexibility in program design and improving information flow from the higher levels of the government to the local government and from the government to the communities.

Government's role as a service provider is very important in disaster risk management capacity. Governments should help communities understand the importance of investing in disaster risk management, provide incentives for improving structural and social vulnerability, and create barriers to shortsighted behavior such as looking for short-term gains by using substandard construction material and illegal construction. The role of change agents, those leading transformational change and linking services to the communities, should be recognized and supported.

7.4 Breaking the Capability Trap

This section focuses on what can be done differently in the Gujarat towns. Building on two radically different approaches, a perception-based joint learning approach is recommended for developing effective capacity on disaster risk management. The first approach is an internal and community-centered approach driven by innovative ways of financing and no reliance on donor funding (Moyo 2009). The second approach is driven by changes in structural design, monitoring, and implementation of donor or "top-down aid" following—"Problem-Driven Iterative Adaptation" approach (Andrews et al. 2013). These two radically different ways of breaking the trap will be discussed and integrated in the context of how capacity development on disaster risk management have occurred in other countries (Table 7.2).

7.4.1 Achieving Community-Centered Disaster Risk Management

The "No Aid" group primarily consists of recent researchers focusing on the review of development aid over the last several decades. The group advocates that the international aid has not been helpful to recipient countries and has in fact reversed the development gains (Birdsall 2007; Moyo 2009; Easterly 2013). Proponents of this group believe that the international aid contributes to the capability trap situation. Moyo (2009), in her book *Dead Aid*, describes how decades of aid have not worked in Africa, have in fact fueled corruption in already-corrupt countries, and are killing national economic growth. Her solution is simple—stop the aid. Relying on trade, attracting foreign investments, and finding local solutions, her proposal is to let the country lead its own development and finances. Before Moyo, William Easterly raised similar points. In his book, the *White Man's Burden*, Easterly (2006) shows how the Western influence and aid has not helped the developing and poor nations. His solution to the problem lies in ensuring a country's own homegrown development based on dynamism of individuals and firms in free markets.

Table 7.2 Comparison of different top-down approaches and the integrated approach

Elements of approach	Small development	Big development	PDIA	PeDJoLA
What drives action?	Internal identification of problems	Externally identified problems or "solutions"—problems identified from deviation from "best practice" forms	Local problem drives development	Similar perception of problems from the government and communities drives development
Planning for action (intervention design)	Incremental planning by individual communities	Advance planning, developing a plan of action	Muddling through—local empowerment for positive deviance	Joint muddling through—local government and community empowerment to solve problems
Implementing action and sustainability	Communities individually implement it on their own	Implementation to be followed as a planned script focusing on disbursement and processes, compliance, monitoring	Problem solving and experimentation with information loops integrated with decisions	Government and community take coordinated actions to resolve common problems, learn, and adapt
			Agents play a key role in implementation and monitoring	Agents play a key role in implementation and monitoring
What results from the action?	Bottom-up—solutions hard to scale up	Top-down—not connected with communities	Diffusion of feasible practice across organizations and communities of practitioners leading to capacity development	Capacity development or trap based on joint learning and empowerment

Source: Adapted from Andrews et al. (2013)

Building on "No Aid" approach, let us first consider the internal and community-centered approach. What would drive individuals and communities in understanding their risk and their strengths in mitigating and preparing for an acceptable level of risks? Can residents work together to identify common solutions and implement? The Gujarat study highlighted crucial questions to start this discussion. Why are citizens of the three Gujarat towns, who know they are at high disaster risk, not taking any actions? Why is a vegetable seller willingly sitting under a crumbling building ready to fall with the slightest tremor? This phenomenon is not typical in Gujarat. Kunreuther et al. (2013) raised similar question looking at citizen behavior in the USA. Just three years after Hurricane Katrina, which killed 1,300 people and caused $150 billion economic damages in Louisiana and Mississippi states of the USA, many residents in Texas refused to follow evacuation warnings during Hurricane Ike leading to deaths of 100.

Why communities and leaders do not take actions, even when the risk is known to them? Box 7.1 below provides some reasons. Rooted in human behavior and

Box 7.1. Why At-Risk Communities and Leaders Do Not Invest in Long-Term Risk Mitigation and Preparedness?

The general understandings on why communities do not take preventive actions are:

Lack of awareness: Citizens might not be aware of the high risk in which they live and so may not invest in risk mitigation measures.

Lack of technical or scientific knowledge: Even if citizens know they are at risk, they might not know what mitigation measures to take.

Budget: An individual may not have adequate budget or financial resources to take mitigation and preparedness actions. Within limited budget, in many low-income and poor families, the trade-off to invest in immediate needs (food and health) over long-term risk mitigation is difficult.

Behavior science, however, provides different explanations rooted in human psychology:

Risk Perception: Citizens perceive risk differently and may underestimate risk for many reasons. Although the reasons for different perceptions are not fully known, the economic status of citizens, the availability to risk information, and the perceived economic benefit of investing in risk mitigation are some of the key ones. The perceived benefit may be less for low-probability event such as an earthquake, even if it is high risk, compared to a frequent event such as investing in a generator to ensure power supply. The psychology of short-term gains over long term is also important in understanding why citizens invest in high-frequency comparatively low-risk event.

(continued)

Box 7.1. (continued)

Sense of powerlessness: Communities might not be empowered to take decisions about their own life and surroundings due to cultural, social, and governance challenges.

Social norms and interdependencies: Social and cultural norms might change a community's behavior. If neighbors are not taking any actions, then justifying spending money on a low-probability event like a major earthquake may not be justifiable for a citizen. If a community leader believes in and takes action, it might help others to follow and take actions.

Learning failure: Learning from a high-stake low probability is different than learning from high-probability low-stake events. Our brain is tuned to learning from high-probability or high-frequency events through trial and error, such as walking and talking. The high-probability low-stake event does not provide time for trial and error.

Disasters bring aid: Many researchers have written about "Samaritan's dilemma"—individuals assume that liberal aid will be forthcoming so why take preventive actions (Kunreuther et al. 2013). A study on Gujarat towns shows that communities invest in developing legal documents for their houses so that they can receive aid. Others have shown that in many instances, drought is declared in areas not affected by drought at all—in anticipation of aid (Sainath 1996). Some other researchers, however, found that communities do not base their decisions on aid expectations.

Disaster declaration helps political agenda: Politicians can benefit from disaster declaration. In the USA, a study looking into a politician's dilemma found that election years are very active time for disaster assistance (Kunreuther et al. 2013). Similarly, focusing on short-term response over long-term risk mitigation also benefits political agenda. Political leaders want to show progress in limited time, leading to rushed projects and quick formation of organizations. Short-term response pays more than long-term risk mitigation.

Government intervention may produce false sense of security: To ensure disaster prevention, it is easier for disaster prevention programs to focus on structural mitigation, such as floodwalls and building retrofitting. The quality of construction and maintenance challenges in a developing country context provide a sense of false security. Individuals will not understand real risk and may lax in taking preventive actions—knowing that they are safe. Even in a developed country context, the example of New Orleans, Louisiana, where floodwall gave away after Hurricane Katrina putting thousands of people at risk, is a good example of such a situation.

Source: Adapted from Kunreuther et al. (2013).

psychology, these reasons point to the importance of understanding how individuals behave or respond to incentives. Bringing change in human behavior is hard to achieve. Communities face difficult challenges—they need to understand social and behavioral biases, de-bias themselves, understand what is good collectively, and invest in long-term mitigation. Social scientists show that communities, on their own, cannot achieve this difficult agenda. Collective action may not necessarily lead to socially optimal solutions due to fee rider problem (Olson 1965). A policy that allows freedom of choice may be socially optimal in the short-term but socially suboptimal in the long-term perspective (Kydland and Prescott 1977). For example, poor residents can settle in low-lying areas of the city as that is the only land available to them. They might think that the more people settle in low-lying areas, the more chances of the government taking actions to legalize their neighborhood and mitigate the flood risk through drainage improvements. In the long run, however, not only such settlements get flooded and cause overflows to nearby neighborhoods, but are also difficult to rehabilitate to safer places. Thus, if residents are collectively not able to make wise choices, governments need to ensure long-term socially optimal perspective (Kunreuther et al. 2013). Regulations and incentives thus remain a crucial part of community-centered risk mitigation.

Some ways governments can strengthen positive behavior in community includes raising awareness, improving risk perception, empowering communities to take charge of their own risks, and providing incentives such as tax incentive for those following building codes for better behavior.

7.4.1.1 Incentives and Strategies for Community-Centered Risk Mitigation

Communicate Current and Future Risks to Residents

- Risk assessment and risk communication: Conducting dynamic risk assessments and making it available to communities will help in joint learning on how disaster, climate, and other risks are changing the region. Land-use plans and zoning ordinances should be based on risk assessment so that prospective buyers can understand the risk of buying a house in a particular location.
- Perception surveys and awareness campaign: Understanding perception and the needs of different communities and members should be a starting point in developing a useful awareness campaign. The local government can lead a massive awareness campaign on developing common perception, understanding risk and risk mitigation activities, and the roles and responsibilities during and after disasters. Education and emergency drills in schools are necessary to achieve preparedness.

Support Positive Behavior

- Tax incentives: For citizens following building codes and land-use plans, property tax incentives can be useful. In the developing country context, however, monitoring implementation of building codes and land-use plans can be a challenge. If building code implementation is encouraged through neighborhood committees, with incentives to award better behavior, collectively there might be a way to resolve the challenge.
- Financial support for strengthening houses and private infrastructure: Financial loans and conditional mortgages for owners undertaking building retrofitting (e.g., seismic or flood proofing) could lead to positive behavior. A partnership with local banks and developers could be useful in this regard.
- Public-private partnership: Involving private business such as shops, banks, and nongovernment organizations in raising awareness and supporting positive behavior can also be beneficial. For example, businesses can be encouraged to adopt earthquake proofing in structures. This can then be showcased to develop public awareness. Ensuring a culture of drills in private sectors may be easy to encourage as it will prevent loss of life of workers.

Keep Communities Involved and in Charge of Their Risk

- Take away false sense of security: While structural mitigation is crucial in ensuring risk reduction, it is important to ensure that the community understands the risk in case structures break up. There are no quick fixes to mitigate risk, and the communities always need to be alert. Awareness and education can help reducing lax attitude.
- Aid is not coming: An awareness campaign that the community is preparing well for disasters and will not need external aid will also help in achieving resiliency. Although politically difficult to achieve, not depending upon external aid can also boost a community's confidence.
- Remember the disaster: Keeping the disaster memory alive is a powerful way of keeping the communities on toes.
- Long-term insurance and long-term mitigation loans: Similar to automobile registration, which requires insurance, premium on housing loans can be reduced with long-term flood or earthquake insurance on a property (Kunreuther et al. 2013). Many cities of developing countries are far away from developing a well-functioning housing insurance market. However, banks can condition their mortgages requiring home owners to purchase insurance if the property is located in a risk zone.

7.4.2 Designing Top-Down Projects Differently

Some critics believe that although internal and community-centered capacity building is possible in theory, in reality it will take a long time, especially in some poor and developing countries. Scaling up the small and locally led development typically is even more difficult falling short of systemic transformation effects (Pritchett et al. 2010). Thus, the second way to enhance capacity is for the donors and governments to undertake ambitious large-scale capacity building efforts. Big development, however, can remain delinked from the communities—offering little on effective capacity at the local level. Big development can lead to a situation where the organizations follow form over function and over time weaken under high expectations (Andrews et al. 2013).

Can countries survive if the international aid is stopped? The "No Aid" group believes that anyway the aid is not reaching needy people and so it is not going to make any difference. Other researchers like Lant Pritchett, Michael Woolcock, and Matt Andrews argue that some countries, if left on their own and continue with their current economic growth rate, will take hundreds of years to reach the level of current per capita income of the USA. They argue to change the way aid is spent in needy countries. Building on Lindbolm's (1959) approach of "muddling through," their solution—Problem-Driven Iterative Adaptation (PDIA)—relies on undertaking incremental reforms to solve the problem, learning from it over time, and accumulating many individual pieces. The approach differs from donor-driven "big project" in many different ways: (i) focusing on problems rather than selling solutions, (ii) creating "authorizing environment" for decision-making, (iii) learning and providing iterative feedback from incremental actions, and (iv) engaging broad sets of agents.

As per the PeDJoLA model of capacity building in disaster risk management and the study of Gujarat towns, capacity building requires working both bottom-up (community) level and top-down (government and donor) level. Bringing the two different perspectives together, we have an integrated way of developing capacity at local level. This approach is contrasted with small and big projects and with PDIA in Table 7.2.

The government has a primary responsibility to ensure the safety and security of citizens. Capacity for managing disasters is thus a key function of government. Three main areas of capacity development are needed at different levels of government for disaster risk management. First and foremost is the prevention and mitigation of disaster risk that includes planning, enforcing, and supervising disaster prevention activities. Local governments and communities need this capacity, with some communities needing behavioral change (as revealed in the study of Gujarat towns) to understand disaster risk and take actions for reducing their risk. Early warning, response, and relief are primary responsibilities of all levels of the government, although many informal community-based institutions play a key role too. Finally, local governments play a key role in recovery. Table 7.3 below summarizes the discussion.

Table 7.3 Components of capacity for disaster risk management

What capacity?	Subcomponents	Level of the government and actors	Citizens
Prevent and mitigate disaster risk	Assessing current beliefs, capacity, and planning	Local	Behavioral change
	Implementation	Local	Behavioral change
	Enforcement and supervision	Local	Behavioral change
Respond quickly and efficiently	Early warning	National, state, local; citizens	Behavioral change, active participation
	Response	National, state, local; informal institutions	Active participation
	Relief	National, state, local; informal institutions	Active participation
Recover quickly and efficiently	Reconstruction	Local; informal institutions	Behavioral change, active participation

Similar to the scope of government and donor support for capacity, as discussed above, timing is also critical to the sustainability and effectiveness of a disaster risk management program. Donors and governments have a big role in correctly utilizing the period immediately after a disaster as a "window of opportunity" to develop and implement effective and pragmatic disaster prevention and preparedness programs. During this period, all major actors are motivated to bring about collective and desired change. As the Gujarat study shows, after the 2001 earthquake in Gujarat, all major actors took actions to fill their perceived capacity deficit gaps. To a certain extent, however, GEERP lost this "window of opportunity" to develop long-term community capacity for disaster prevention because it did not focus on improving local communities' fundamental attitudes and understanding about disasters.

7.5 International Experience in Developing Effective Disaster Risk Management Capacity

International experience shows that there is no "one size fits all" for steps to develop capacity on disaster risk management. Most countries are constantly evolving their approach to better manage disasters. Examples of Chile and Turkey show the importance of continuously investing in disaster risk management. Due to investments in risk mitigation and preparedness, Chile suffered lesser impacts from the 2010 earthquake, which was of higher impact compared to an earthquake in Haiti in the same year (see Box 7.2). Turkey learned from the 1999 Marmara earthquake and radically changed its policies and institutions related to disaster risk management (see Box 7.3). The response to Cyclone Phailin in the eastern coastal regions of India in 2013 further demonstrates the importance of steadily learning from disasters and investing in preparedness (see Box 7.4).

Box 7.2. Why Chile Suffered Less Impact from the 2010 Earthquake?
Chile and Haiti both suffered strong earthquakes in 2010. Chile's earthquake (8.8 Mw) was 500 times stronger than that of Haiti (7Mw) (USGS 2010). Compared to Haiti, however, Chile suffered relatively less impacts. The death count in Chile was 526 with disaster damages and losses equivalent to 18 % of country's GDP (US$30 billion) (Fermandois 2011). Haiti's 2010 earthquake killed 200,000 and led to a total damages and losses equivalent to slightly more than the country's total GDP in 2009 (US$7.8 billion) (Government of Haiti 2010). Apart from being economically better off (as higher middle-income country), what was different in Chile compared to Haiti?

The huge difference in the impacts in the two countries is due to long-time investment in disaster preparedness and structural strength of infrastructure and buildings. Chile invested in an earthquake risk management system and mandated earthquake proofing for new structures, requiring that materials like rubber and features like counterweights be built into the architectural designs to allow buildings to bend and sway rather than break during temblors. This preparedness stemmed from active government involvement in strengthening national infrastructure and in creating and enforcing strict regulations (Freeman et al. 2003). In Haiti, planning, disaster preparedness, and earthquake-resistant building codes were nonexistent or outdated.

A closer look at Chile shows a comprehensive disaster risk management system, which has evolved after the 1965 earthquake with the setting up of the Office of National Emergencies (*Oficina Nacional de Emergencia*, ONEMI) under the Ministry of Interior. Slowly, the system has moved from emergency response toward risk mitigation and prevention. Although it appears centralized and hierarchical, the Chilean system is composed of committees at the community, provincial, and regional levels that are responsible for evaluating proposed actions and designing and prioritizing prevention, mitigation, and preparedness projects appropriate to each administrative level. In the case of an emergency, all of the available resources in the affected community are used first. If the magnitude of the event exceeds the local capacity, additional resources are mobilized successively from the provincial, regional, and national levels (Freeman et al. 2003).

Following the 2010 earthquake, the government is putting even more effort in adopting public policies that incorporate risk reduction in all sectors. Chile is now recognizing the importance of decentralizing implementation, promoting community empowerment, and respecting local identities in risk reduction actions (IBRD 2012).

Box 7.3. Evolution of Disaster Risk Management System in Turkey

The 1999 Marmara earthquake (Kocaeli and Düzce) was a turning point in Turkey's disaster risk management history. The earthquake killed 17,000 people, made 100,000 residents homeless, and caused $5 billion in damages and losses. The country, which is historically prone to earthquakes—with the first recorded earthquake in the year 325—started a process of critically revaluating disaster management in 1999 after the 45 seconds long earthquake. A rebirth of the system occurred with the establishment of a General Directorate for Disaster Management, working with the Prime Ministry; inauguration of a mandatory disaster insurance program; update in the seismic design codes; and improvement in the regulations for search and rescue. Formal disaster management training and education was introduced following this major earthquake.

The disaster management system in the country, however, remains centralized and hierarchical. The Turkish Emergency Management Agency is in charge of risk reduction and preparedness at the national level. At a provincial level, the Provincial Governor is in charge of disaster response plans and mitigation—which can seek help from national agency in the case of a big disaster. At a local level, the municipality has city planning and infrastructure development functions. The city of Istanbul has been a leader in risk mitigation with the formation of the Istanbul Emergency Management Agency, AKOM.

The government has worked with many international partners such as the World Bank in post-earthquake reconstruction and building retrofitting programs. The Istanbul Seismic Risk Mitigation and Emergency Preparedness Project (ISMEP), a $400 million loan to the government, focuses largely on retrofitting of public buildings and on developing emergency information system in the city. Similar to the Gujarat case, the coordination with different government agencies at all levels and better involvement of communities still remain a challenge in the country.

Source: Ural (2012)

Box 7.4. Steadily Learning from Disasters: Timely Warning and Evacuations Saved Thousands in Cyclone Phailin Response in India

Compared to the 1999 cyclone, which struck the eastern coast of India and killed close to 10,000 people, mortality from Cyclone Phailin—which was of similar intensity and struck similar areas of India in 2013—was a mere 38 (DevPolicy Blog 2013). The difference in mortality rate was primarily contributed to effective storm warnings issued by the Indian Meteorological Department and India's largest storm evacuation ever—relocating 900,000 people from the coastal areas to shelters.

(continued)

Box 7.4. (continued)

Some news reports talked about potential weakening of the storm before landing. Nevertheless, timely warning and evacuation remain the most extraordinary response to Cyclone Phalin (Washington Post 2013). The results were celebrated as the outcome of over ten years of collaboration between the government, communities, and international partners. Over the past ten years, the government, aid agencies, and communities have worked together to ensure plans in place for disaster-prone areas. Activities such as disaster simulations, hazard mapping, and improving community resilience, combined with improvements in technology—such as mobile phones and weather tracking systems—in high-risk areas have ensured the government and communities are ready when emergency strikes. Communities knew far earlier of the oncoming storm and the authorities were able to provide clear instructions about what to do.

In highly advanced and high at-risk countries, disaster risk management started as centralized and hierarchical systems and is now moving toward focusing on local areas and enhancing community resilience. These countries have invested in incrementally learning after deadly disasters, changing laws, institutions, and approaches to managing disaster impacts and mitigating risks. The US and Japan cases demonstrate the above points (see Boxes 7.5 and 7.6). Both countries have invested heavily in developing preparedness and mitigation programs. They have comprehensive and complex national emergency management programs, which have evolved over the years in response to numerous disasters. The programs remain highly complex with both countries focusing on developing public-private partnerships on sharing disaster losses. In Japan, earthquake risk insurance is offered by private insurers as a part of fire insurance policies. In the USA, a similar but importantly different public-private partnership exists to cover flood losses. In the USA, the challenges are related to the loss sharing and mitigation actions that have instilled false sense of security such as structural measures (as seen in Hurricane Katrina) or flood insurance (which has made it easier to buy houses in high-flood-risk areas). Japan, on the other hand, has the most developed earthquake monitoring and response systems but still struggles to cope up with increased and multiple hazard risks. In both countries, the new research is now supporting community resilience.

7.6 What Can Be Done Differently? A Program for Gujarat Towns

To develop effective capacity on disaster risk management at societal level, primary focus on local and community resilience should be given. Following the new model, this would involve perception-based joint learning at the local level—building capacity and ownership in local government and community.

Box 7.5. USA: Facing Recurrent and Costly Disasters

In the USA, disaster spending has increased (Weiss and Weidman 2013) over the last 30 years, primarily on relief and rescue. Just one event, Superstorm Sandy in 2012, caused $50 billion in damages and $60 billion relief and recovery.

A complex disaster management system has evolved in the country following major disasters. The Federal Emergency Management Agency (FEMA) is responsible to set policy goals and provide funding to carry out emergency management objectives. FEMA also helps to train emergency management personnel, approves state and local Hazard Mitigation Action Plans, and oversees the distribution of recovery assistance. Each state has an appointed emergency management office. Usually, these departments are located within the governor's office or the department of public safety. They may also be standalone agencies or integrated into state homeland security. Local governments (county and municipal) have emergency management offices. FEMA works with many government agencies as well as communities and citizens to ensure preparedness. Risk mitigation activities are mostly dealt by specialized agencies with states and localities having ultimate decision-making power.

A number of challenges exist in the country to achieve effective disaster risk management capacity at all levels. First, FEMA and emergency management structure remain centralized, while the growing hazard requires a different approach focusing on enhancing resilience in a changing climate (Weiss and Weidman 2013; McEntire 2012). Instead of focusing entirely on emergency operation plans, the country needs to find ways to reduce vulnerability and enhance capabilities. Second, regional cooperation and coordination among all the actors involved in emergency management in the United States need to be improved.

To meet the challenges, social scientists have urged to invest more in behavioral change of communities. By rewarding positive behavior, the new approach needs to focus on encouraging longer-term risk reduction through mortgage loans, tax incentives, and other individual-based incentives (Kunreuther et al. 2013). Others have called for an enhanced focus on community resilience, by setting community resilience fund for local disaster mitigation and preparedness (Weiss and Weidman 2013).

Box 7.6. Japan: From Readiness to Resilience

Perched on the Ring of Fire, an arc of seismic activity that encircles the Pacific Basin, Japan is one of the most earthquake-prone countries in the world—and it is also one of the best equipped to handle them. Having survived the quake of 1923, the utter devastation of World War II, and, later in 1995, the earthquake in Kobe, the country has done more than most countries when it comes to disaster preparedness (Time Magazine 2011). Japan boasts the world's most sophisticated earthquake early warning systems. Emergency drills organized by public and private organizations work, among other things, to transport "stranded" commuters from their offices to their homes. Japan's tsunami warning service, set up in 1952, consists of 300 sensors around the archipelago, including 80 aquatic sensors that monitor seismic activity 24/7. The network is designed to predict the height, speed, location, and arrival time of any tsunami heading for the Japanese coast. Tsunami safety has been a focus of coastal city planning throughout the nation. On Japan's east coast where tsunamis frequently hit, hundreds of earthquake and tsunami-proof shelters have been built. Some cities have built tsunami walls and floodgates so that the waves don't travel inland through river systems.

Japan has a very complex institutional setup for disaster management, which has evolved after each major disasters and accidents. The current setup came into being after severe damages from Typhoon Isewan in 1956, which led to the adoption of the Disaster Countermeasures Basic Act in 1960. Under this act, the Central Disaster Management Council was formed with the Prime Minister as the chairperson and ministers of all departments, heads of major public institutions, and experts as members. Within the Cabinet Office, which is the secretariat for this Council, the Minister of State for Disaster Management is responsible for planning and central coordination on disaster risk reduction. In prefectures and local municipalities, the prefectural and municipal Disaster Management Councils are established with the members of representatives of local government organizations including police and fire management departments as well as designated local public corporations. Implementation of disaster risk management measures is based on the local disaster management plans drafted by the Councils.

After the adoption of the Disaster Countermeasures Basic Act, the overall trend for disaster impacts went down till the Awaji earthquake in 1991 and more recently the Tohoku earthquake and tsunami. Changes in the system were introduced after these two major events, with the emphasis on ensuring resilient communities. The Government Committee for Technical Investigation on Countermeasures for Earthquakes and Tsunamis after the Tohoku earthquake, apart from improving assumptions about severity and scale of hazards, potential damage scenarios, and risk mitigation actions, emphasized the importance of investing in resilient communities. These include ensuring that memory of historical disasters remains in communities (with simple historical stone edicts), risk mitigation through land-use planning and building codes, considering community characteristics in disaster mitigation programs, promoting women in local Disaster Management Councils, and raising awareness.

Source: Adapted from Cabinet Office of Government of Japan (2011)

7.6.1 Enhancing Government Capacity

The international experiences show that a majority of the countries started their journey for disaster risk management through policy and institutional changes after major disasters. The case of Gujarat is no exception. The establishment of the GSDMA and its activities gives a solid foundation to build upon. The GSDMA has developed an organizational structure that has a presence in both urban and rural areas. The GSDMA was able to coordinate post-disaster reconstruction in all disaster-affected towns, has received various awards and recognition for its work, and continues to be a role model to other disaster risk management agencies across the developing world. It has dedicated state funding for its growing operations and is improving efforts to develop and maintain early warning systems, risk assessments systems, training, and awareness-building programs.

Compared to the situation just after the earthquake in 2001, there is a substantial change in the three study towns following the introduction of the GEERP. At an administrative level, there are profound changes in government capacity to manage disasters. Starting with the setting up of a new organization to respond to disasters and manage disaster risk, the state administration's efforts have shown results, legitimacy, and resilience.

The GSDMA's structure could be used to include more citizens in the development and implementation of disaster prevention programs and plans. There is interest among the government staff in increasing citizen participation. However, agencies will need more support to carry out this process. Local governments need to be empowered to make decisions regarding land use and zoning. They need more resources either through national and state governments or through public-private partnership to undertake their role effectively.

7.6.2 Developing Community Capacity

At a social level, awareness about risks and preparedness for disasters are an important aspect of capacity building. The training and awareness programs developed by the government are the primary sources of involvement in the communities. Many people keep the past experiences alive in their memory. This could have a strong impact on social capacity to deal with future disasters. Citizens' possession of various lifesaving skills and preparedness skills is important in building social capacity. These are, however, missing in the study towns. Many local masons received training on safe building techniques. This was initiated by the GSDMA and the training was provided by NGOs who had expertise in this field. These training programs were offered during the reconstruction and rehabilitation phases but there have been no training programs to refresh the skills or to add more people with these skills. Since the 2001 earthquake, the skills of some individuals, especially men, have increased. However, the number of individuals who effectively gained new skills is very low and the skills are getting lost with time. The development of new construction techniques (e.g., concrete and masonry) that replace the old techniques (e.g., stone and mud construction) has also made some of the traditional mason skills obsolete. Disaster risks are not better understood at the local level and continue to increase with illegal construction, masons losing their skills related to safer building construction techniques, and lack of coordination. Strikingly, there is no interest from the citizens to keep the focus on disaster risk management.

7.6.3 Achieving Joint Perception-Based Learning

7.6.3.1 Joint Perception Survey, Risk Understanding, and Learning

Government and communities jointly need to understand their risk landscape, how current actions are changing future risk, and adapt. GSDMA has conducted risk assessments. These need to be available at local levels along with mechanisms to update them periodically to capture updated risk mapping. A massive public awareness campaign to develop a common understanding of the risks, causes, and impacts of disasters will be helpful in developing common perception and need for risk mitigation measures. Ideally, a small disaster memorial in the city can house risk maps (in digital or paper format) for communities to see their risk to multi-hazards. Government actions and results also need to be displayed to get a higher buy-in.

7.6.3.2 Risk-Based Spatial Planning, Zoning, and Building Codes

The empowered local governments, with better budget and staff, need to undertake risk-based spatial planning and zoning. The building codes are already updated in the study towns but need to be reviewed for multi-hazard risks. Risk communication through zoning is important. If an area is on a low-lying part of the city or near a fault line, zoning should be able to say that. In the context of Gujarat towns, enforcing building codes and zoning is a big challenge.

7.6.3.3 Involvement, Empowerment, and Ownership

Involve community groups in planning and implementing projects. Form community groups for disaster preparedness with members coming from women, elderly, and all classes/castes. Developing public-private partnership in encouraging such groups is essential. The groups can also be made responsible for providing training on building construction techniques, implementing building codes, and retrofitting measures. Financial support can be provided for the families (based on income group) willing to take such initiative under the community group's monitoring.

7.6.3.4 Incentives for Positive Behavior

Property tax incentives and long-term mortgages in collaboration with banks and real estate developers can provide powerful incentive to improve building code enforcement. Public-private partnership in rewarding good behavior—those participating in awareness campaign, improving building construction, and actively involved in preparedness—can encourage citizens to keep taking positive steps. Periodic joint learning to remember and learn from the disasters should also be undertaken. Agents—political leaders, private sectors, and CBOs—should be involved and appreciated in such events.

Thus, to ensure long-term capacity development for disaster risk management, the governments need to invest in (i) empowering local government agencies, municipalities, and development authorities to lead the disaster risk management agenda by providing technical, financial, and decision-making power to design and implement public awareness programs, enforcement of building codes, contingency planning, coordination, and risk-sensitive land-use planning; (ii) massive public awareness campaigns to develop a common understanding of the causes and impacts of disasters; (iii) engaging citizens, NGOs, and CBOs for maintaining contingency plans and providing incentives such as tax rebate or housing loan subsidy if a new house is built earthquake resistant for encouraging behavioral change; and (iv) supporting the critical roles of local government officers and community leaders—the change agents—by rewarding them for their innovative practice and dedication.

References

Andrews, M. (2008a). Good government means different things in different countries. *Governance, 23*(1), 7–35.

Andrews, M. (2008b). *Are one-best-way models of effective government suitable for developing countries?* Faculty Research Working Papers Series RWP08-014, John F. Kennedy School of Government, Harvard University, Boston.

Andrews, M., Pritchett, L., & Woolcock, M. (2013). Escaping capability traps through problem driven iterative adaptation (PDIA). *World Development, 51*, 234–244.

Beck, U. (2006). Living in the world risk society. *Economy and Society, 35*(3), 329–345.

Birdsall, N. (2007). Do no harm: Aid, weak institutions and the missing middle in Africa. *Development Policy Review, 25*(5), 575–598.

Brinkerhoff, D. W. (2004, May 12–13). *The enabling environment for implementing the millennium development goals: Government actions to support NGOs.* Paper presented at George Washington University Conference "The Role of NGOs in Implementing the Millennium Development Goals", Washington, DC.

Brinkerhoff, D. W. (2008). The state and international development management: Shifting tides, changing boundaries, and future directions. *Public Administration Review, 68*(6), 985–1002.

Cabinet Office of Government of Japan. (2011). http://www.bousai.go.jp/1info/pdf/saigaipanf_e.pdf. Accessed Apr 2014.

Cardona, O. D. (2003). The need for rethinking the concepts of vulnerability and risk from a holistic perspective: A necessary review and criticism for effective risk management. In G. Frerks, G. Bankoff, & D. Hilhorst (Eds.), *Mapping vulnerability: Disasters, development, and people.* London: Earthscan.

Corbacioglu, S., & Kapucu, N. (2006). Organizational learning and self-adaptation in dynamic disaster environments. *Disasters, 30*(2), 212–233.

DevPolicy Blog. (2013). *Being prepared: India's response to Cyclone Phailin*. http://devpolicy.org/being-prepared-indias-response-to-cyclone-phailin-20131024/. Accessed Apr 2014.

Easterly, W. R. (2006). *The white man's burden: Why the West's efforts to aid the rest have done so much ill and so little good*. New York: Penguin Press.

Easterly, W. (2008). Institutions: Top down or bottom up? *American Economic Review, 98*(2), 95–99.

Easterly, W. R. (2013). *The tyranny of experts: Economists, dictators, and the forgotten rights of the poor*. New York: Basic Books.

Fermandois, A. (2011). *Chile and its earthquake – Preparedness, response and lessons*. http://dels.nas.edu/resources/static-assets/materials-based-on-reports/presentations/AmbassadorFermandois.pdf. Accessed June 2014.

Freeman, P. K., Martin, L. A., Linneroot-Bayer, J., Mechler, R., Pflug, G., & Warner, K. (2003). *National systems for the comprehensive management of disaster risk: Financial strategies for natural disaster reconstruction*. Washington, DC: Inter-American Development Bank.

Government of Haiti. (2010). *Haiti earthquake PDNA: Assessment of damages, losses, general and sectoral needs*. http://www-wds.worldbank.org/external/default/WDSContentServer/WDSP/IB/2012/06/19/000333038_20120619012320/Rendered/PDF/701020ESW0P1190R0Haiti0PDNA020100EN.pdf. Accessed June 2014.

Grindle, M. S. (2007). Good enough governance revisited. *Development Policy Review, 25*(5), 553–574.

IBRD. (2012). *Improving the assessment of disaster risk to strengthen financial resilience. A special G20 publication by the government of Mexico and World Bank*. Washington, DC: International Bank for Reconstruction and Development.

Ingram, J. C., Franko, G., Rumbaitis-del Rio, C., & Khazai, B. (2006). Post-disaster recovery dilemmas: Challenges in balancing short-term and long-term needs for vulnerability reduction. *Environmental Science & Policy, 9*(7–8), 607–613.

Jha, A., Bloch, R., & Lamond, J. (2012). *Cities and flooding: A guide to integrated urban flood risk management for the 21st century*. Washington, DC: The World Bank.

Jigyasu, R. (2002, May 23rd to 25th). *From Marathwada to Gujarat. Emerging challenges in post-earthquake rehabilitation for sustainable eco-development in South Asia*. Proceedings of the conference improving post-disaster reconstruction in developing countries. University of Montreal, Montreal, Canada. Available at: www.grif.umontreal.ca/pages/i-rec%20papers/rohit.PDF. Accessed 16 Sept 2012.

Kunreuther, H., Meyer, R., & Michel-Kerjan, E. (2013). Overcoming decision biases to reduce losses from natural catastrophes. In E. Shafir (Ed.), *The behavioral foundations of public policy*. Princeton: Princeton University Press.

Kydland, F., & Prescott, E. (1977). Rules rather than discretion: The inconsistency of optimal plans. *Journal of Political Economy, 85*, 473–491.

Lindbolm, C. E. (1959). The science of muddling through. *Public Administration Review, 19*(2), 79–88.

McEntire, D. A. (2012). Emergency management in the United States: Disasters experienced, lessons learned, and recommendations for the future. In McEntire, D. A. (Ed.), *Comparative emergency management: Understanding disaster policies, organizations, and initiatives from around the world*. Washington, DC: FEMA.

Mileti, D. S., & Fitzpatrick, C. (1993). *The great earthquake experiment: Risk communication and public action*. San Francisco: Westview Press.

Moyo, D. (2009). *Dead aid: Why aid is not working and how there is a better way for Africa*. New York: Farrar, Straus and Giroux.

Olson, M. (1965). *The logic of collective action*. Cambridge, MA: Harvard University Press.

Pritchett, L., & Weijer, F. (2010). *Fragile states: Stuck in a capability trap?* Input paper for The World Development Report 2011. Washington, DC: The World Bank.

Pritchett, L., Woolcock, M., & Andrews, M. (2010). *Capability traps: The mechanisms of persistent implementation failure*. Working Paper 234. Washington, DC: Center for Global Development.

Sainath, P. (1996). *Everybody loves a good drought: Stories from India's poorest districts*. New Delhi: Penguin.

Sengezer, B., & Koc, E. (2005). A critical analysis of earthquakes and urban planning in Turkey. *Disasters, 29*(2), 171–194.

Time Magazine. (2011). *How Japan became a leader in disaster preparation*. http://content.time.com/time/world/article/0,8599,2058390,00.html. Accessed Apr 2014.

UNESCO. (2009). *Guidebook for planning education in emergencies and reconstruction*. Paris: UNESCO.

Ural, D. N. (2012). Emergency management in Turkey: Disasters experienced, lessons learned, and recommendations for the future. In McEntire, D. A. (Ed.), *Comparative emergency management: Understanding disaster policies, organizations, and initiatives from around the world*. Washington, DC: FEMA.

USGS. (2010). *The MW 7.0 Haiti earthquake of January 12, 2010: USGS/EERI advance reconnaissance team report*. http://pubs.usgs.gov/of/2010/1048/of2010-1048.pdf. Accessed May 2014.

Washington Post. (2013). *Major disaster averted: 5 reasons why cyclone Phailin not as bad as feared in India*. http://www.washingtonpost.com/blogs/capital-weather-gang/wp/2013/10/14/major-disaster-averted-5-reasons-why-cyclone-phailin-not-as-bad-as-feared-in-india/. Accessed Apr 2014.

Weiss, D. J., & Weidman, J. (2013). *Disastrous spending: Federal disaster-relief expenditures rise amid more extreme weather*. Center for American Progress. http://www.americanprogress.org/issues/green/report/2013/04/29/61633/disastrous-spending-federal-disaster-relief-expenditures-rise-amid-more-extreme-weather/. Accessed 6 Apr 2014.

Wildavsky, A. B. (1988). *Searching for safety*. New Brunswick: Transaction.

Winchester, P. (2000). Cyclone mitigation, resource allocation and post-disaster reconstruction in South India: Lessons from two decades of research. *Disasters, 24*(1), 18–37.

Wisner, B., Blaikie, P., Cannon, T., & Davis, I. (2004). *At risk: Natural hazards, people's vulnerability, and disaster*. New York: Routledge.

World Bank. (2005). *Capacity building in Africa: An OED evaluation of World Bank support*. Washington, DC: The World Bank.

World Bank. (2011). *Enabling environment for civic engagement*. Available at: http://go.worldbank.org/4JJJ1MKSZ0. Accessed 12 May 2011.

Chapter 8
From Capability Trap to Effective Disaster Risk Management Capacity: What Can Governments, Communities, and Donors Do?

Abstract This chapter focuses on understanding what the future holds for disaster risk management. The role of governments and communities in developing effective disaster risk management capacity is explored along with recommendations. The chapter focuses on two interrelated topics. First, do donors and international aid have a role to play? Second, how much capacity development for disaster management is actually about making the development happen in a right manner? This research has helped in conceptualizing how capacity building takes place and what factors are important for its effectiveness and sustainability at a local level. With a better understanding of determinants of capacity development, scarce resources can be effectively allocated. The findings will be useful in designing and implementing effective disaster risk mitigation programs that keep people and their assets safe from disasters, relying on better engagement with community, and supporting the critical role of change agents. This final chapter also discusses some methodological limitations and offers directions for future research.

Keywords Effective capacity • Actors • Role of donors • Change agents • Risk-sensitive development

Returning to the questions raised in Chap. 1, this chapter focuses on understanding what the future holds for disaster risk management. Will disaster impact continue to grow? How can effective disaster risk management capacity be developed? The role of governments and communities in developing effective disaster risk management capacity is explored along with recommendations. The chapter focuses on two interrelated topics. First, do donors and international aid have a role to play? Second, how much capacity development for disaster management is actually about making the development happen in a right manner?

This overall research has helped in conceptualizing how capacity building takes place and what factors are important for its effectiveness and sustainability at local level. With a better understanding of determinants of capacity development, scarce resources can be effectively allocated. The findings will be useful in designing and implementing effective disaster risk mitigation programs that keep people and their assets safe from disasters, relying on better engagement with community, and

A. Tiwari, *The Capacity Crisis in Disaster Risk Management*,
Environmental Hazards, DOI 10.1007/978-3-319-09405-2_8

supporting the critical role of change agents. This final chapter also discusses some methodological limitations and offers directions for future research.

8.1 Disaster Impacts Will Increase

If current business as usual continues, disaster impacts will increase. Global- and country-level assessments support this trend. Study in Gujarat towns as well as experience in many countries show that, even with massive investments, there is a long way to develop effective disaster risk management capacity. Escalating losses are the result of economic and behavioral forces (Kunreuther et al. 2013). On the economic side, rising exposure of people and economic assets to hazards will drive economic damages and losses from disasters. As the world becomes crowded with cities—with more people and assets accumulating in smaller areas—and as the planet warms bringing more severe and frequent hazards (IPCC 2012), the trend for disaster losses is clear. On the behavioral side, both decision-makers and citizens are not able to attend to low-probability and high-impact events as they focus on high-frequency low-impact events and short-term planning horizon (Kunreuther et al. 2013). Decision-makers focus on short-term gains, while residents do not see the benefit of investing in an event that seems unlikely in their lifetime. Once the catastrophic event happens, however, the results are dramatic.

In a developing country context, the capacity to reverse the trend is even more limited as the decision-makers are stretched thin for making investments in many sectors in response to the growing population. In addition, incentives to invest in short-term gains remain high. Thus, institutions that follow "form" are easy to create than to ensure their function. One-time investment on structural mitigation and training are easier than long-term risk mitigation and ensuring learning. Meeting numerical targets for improving structures is easier than ensuring higher community involvement and empowerment. The result is capability trap, leading to higher and escalating disaster losses.

8.2 Effective Capacity Can Be Developed for Reducing Disaster Impacts

On the positive side, the global community is giving increasing attention to disaster risk management. The Hyogo Framework for Action (UNISDR 2007) and Intergovernmental Panel on Climate Change are key international frameworks supporting disaster risk management and climate change adaptation. As the model presented in this book shows, the effective capacity includes government and community capacity, primarily at the local level (where disaster impacts are highly felt).

8.2.1 *Actors and Elements of Effective Capacity for Disaster Risk Management*

The conceptual model presented in Fig. 5.4 shows the basic framework for understanding capacity development for disaster risk management. The model and fieldwork show how to develop capacity through a locally transformative process of change and how to define capacity development and capability trap (see Table 8.1 below). Both government and community institutions including NGOs, CBOs, and private sector need to work together to be able to ensure joint learning, resilience, and adaptability. In the Gujarat study, current capacity or capability trap was measured through various indicators developed for capacity and sustainability. Capacity of both government and nongovernment institutions can be increased by supporting governments' role as a service provider, ensuring that the local government and the community are empowered, and supporting change agents.

Various actors have key roles to play in developing and sustaining capacity for managing disasters. Governments have the responsibility to design capacity development interventions. Such interventions should start with an assessment of the environment, the system and disaster risk, as well as the community's vulnerability, beliefs, perceptions, and acceptable level of risk. Local governments and communities should be at the forefront of capacity development. Donors bring significant resources and have a key interest in investing in disaster prevention. Rather

Table 8.1 Components of effective capacity for disaster risk management

Actors	Capacity development	Capability trap	Indicators	To increase capacity
Government	Results, legitimacy, resilience	Failure, lack of recognition, rigid structure and function	Enabling law funding, staff, technical ability, political priority	Laws, budget, technical staff, government's role as a service provider
Communities and civil society organizations	Behavioral change, positive attitude—valuing life and property, taking control of the future	No control over future, helplessness, apathy, disregard to social context, disintegrated actions/demands	Community priority, behavior change, empowerment	Be more aware of risks, take informed decisions, and demand better prevention
Both (locally transformative process of change)	Joint learning, resilience, adaptability, community empowerment	Negative changes after leadership changes, rigid programs—small or very big which have not taken roots in the social systems; reliance on donors	If both government and nongovernment agents work together	Campaigns, use of social capital, incentives for change agents

than supporting top-down and supply-driven capacity development programs, they should support joint capacity building programs resulting from consultations and cooperation with communities. The framework should include support for the environment and change agents. Nongovernmental organizations such as NGOs, CBOs, community groups, the private sector, and the media need to be proactive in asking the government to involve the community and to coordinate better. Citizens have a major role, especially in a democracy, to ask for better services and responses. They should realize their role in ensuring coordination and demanding information.

8.2.2 Role of Donors

Though the "No Aid" group argues not to provide donor aid to developing and poor countries, the donors will be investing in disaster response and mitigation given escalating disaster impacts. With their role in providing funding and technical knowledge, they can play a crucial role in ensuring effective capacity development. Thus, the key question is not whether they have a role to play but what should they finance. Their role should be focusing on five major principles:

1. *Focus on long-term risk mitigation measures*: Donors need to be candid about the long time frame for developing disaster risk management capacity along with the need for long-term financing and support.
2. *Invest in global risk assessment, risk indices, risk communication, and monitoring*: Just like local governments in cities and towns, at an international level, donors can support investment in unified global risk assessment and risk indices.[1] Communicating risk and monitoring them over time will also be crucial in understanding changing global risk landscape.
3. *Technical leadership*: Donors can provide technical leadership in suggesting optimum solution for building long-term risk mitigation capacity. As the new model in this book suggests, such capacity needs to rely on building long-term risk mitigation and providing incentives for positive behavior.
4. *Put the recipient country in driver's seat and finance projects that focus on developing local and community capacity*: At national level, donor support can focus on ensuring mortality reduction in developing countries through investment in early warning and response and steps taken to mainstream long-term risk mitigation in development. Bulk of investment, however, should focus on local-level capacity development and community-level behavioral changes.
5. *Emergency response and post-disaster reconstruction need to include long-term risk mitigation*: Donor support to emergency response should include support for long-term risk mitigation.

[1] Global Assessment Report and IPCC are currently such mechanisms but adopting a more unified approach with clear global risk indicators can be useful.

8.2.3 Ensuring Risk-Sensitive Development

Ensuring that development includes ways of mitigating long-term disaster risks is at the heart of reducing disaster losses. Disaster risks emerge from and exacerbate from skewed development processes such as those associated with environmental degradation, rapid and unplanned urbanization in hazardous areas, failures of governance, and the scarcity of livelihood options for the poor (IPCC 2012). Effectively, disaster risk management should focus on ensuring disaster risk reduction in national development and sector plans (IPCC 2012). This should include constantly analyzing risks, not just physical hazards but financial and social risk, and taking appropriate risk mitigation measures in development planning (World Bank 2013). Sectoral ministries need to consider risks in their sectors and take appropriate actions. Prime focus should, however, be on local level where empowerment and inclusiveness can provide wider and long-term benefit. After all, development is about not getting stuck in capability trap but developing effective capacity.

8.3 Contributions and Remaining Questions

The materials presented in this book are one of the first attempts at bringing together different strands of the capacity development paradigm as they apply to the disaster risk management field. The integrated model developed for this research can be a starting point for future research to understand, design, and analyze capacity building programs. It not only will help the government of Gujarat in improving its disaster risk management program but can also help other governments and international development agencies in developing future programs. The successful and sustainable governance aspects of the framework can provide guidance in developing public policies that integrate the top-down and bottom-up aspects and perceptions. With a better understanding of determinants of capacity development, scarce resources can be more effectively allocated and employed.

8.4 Directions for Future Research

This book systematically presented capacity development within a complete framework that has rarely been attempted by researchers before. Prior research looked primarily at top-down characteristics. This book, while validating the top-down characteristics, brought out the importance of a bottom-up perspective. It also looked at the qualitative aspects of capacity development, thus going beyond such quantitative aspects as how many additional staff were recruited or how many training programs were conducted. The book looks at the sustainability factors of capacity development that have rarely been attempted in earlier research.

The model presented in this book needs to be applied in different national settings to identify other factors of influence. Comparative studies across countries could provide opportunities to compare and contrast the capacity development factors as well as sustainability factors. New research can apply the conceptual framework to other geographical areas, such as cities in other developing countries that have been affected by disasters. The framework can also be applied to other kinds of disasters such as floods, tsunamis, landslides, and hurricanes. Within India itself, capacity building under GEERP can be compared with disaster reconstruction programs undertaken in other states, particularly in the states of Orissa and Tamil Nadu. The national government has started encouraging states to develop institutions that can comprehensively address disaster management. Analyses of national policies would be another avenue for follow-up.

Many current studies evaluate different aspects of capacity building in a piecemeal fashion. The conceptual framework proposed in the Gujarat study provides a holistic way of understanding capacity development; however, this research did not focus on understanding how capacity development interventions were implemented under GEERP, which is important if we are to understand how capacity building takes place on the ground from agenda setting to design, adoption, and implementation (Grindle 2007). Future research may wish to explore this aspect in more detail, focusing on understanding how programs were implemented.

Similarly, while indicators for measuring government capacity were comprehensive for this research, indicators for community capacity need more elaboration as they do not capture networking and social capital analysis necessary for disaster risk management. Future research may wish to develop further indicators for assessing community capacity, including indicators for capturing how integrated the government and community capacities are. New research can also focus on defining environmental and system factors (context as defined by Grindle 2007) as well as the capacity building potential and capability traps of various groups and subgroups that work in formal (government system) and informal (community) settings. A retrospective study of other programs at different time frames might be another avenue of future research.

Extensive data such as time series data and panel data of different towns could be collected through surveys or from government records. Various statistical analyses could be employed to analyze these datasets. Other data collection methods such as focus group discussions could be carried out to determine whether there are consistent findings. Surveys and interviews of a larger sample size could bring out different or much more elaborate findings. Future research may also wish to employ interviews and other methods with communities in Bhachau and Mandvi to examine if the findings are consistent with this research. However, the cost of collecting such extensive data, both in terms of financial cost and time required, needs to be compared with the additional benefit from the potential data.

References

Grindle, M. S. (2007). Good enough governance revisited. *Development Policy Review, 25*(5), 553–574.

IPCC. (2012). Managing the risks of extreme events and disasters to advance climate change adaptation. In C. B. Field, V. Barros, T. F. Stocker, D. Qin, D. J. Dokken, K. L. Ebi, M. D. Mastrandrea, K. J. Mach, G.-K. Plattner, S. K. Allen, M. Tignor, & P. M. Midgley (Eds.), *A special report of working groups I and II of the Intergovernmental Panel on Climate Change*. Cambridge, UK/New York: Cambridge University Press. 582 pp.

Kunreuther, H., Meyer, R., & Michel-Kerjan, E. (2013). Overcoming decision biases to reduce losses from natural catastrophes. In E. Shafir (Ed.), *The behavioral foundations of public policy*. Princeton: Princeton University Press.

UNISDR. (2007). *Hyogo framework for action (2005–2015) – Building the resilience of nations and communities to disasters*. Extract from the final report of the World Conference on Disaster Reduction (A/CONF.206/6). Geneva. http://www.unisdr.org/files/1037_hyogoframeworkforactionenglish.pdf

World Bank. (2013). *World development report 2014*. Washington, DC: The World Bank.

Appendix: Interview and Survey Sampling

Interview Sample

Interviews were carried out with senior managers/staff and top elected officials at government agencies, municipalities, and other organizations in Bhuj, Bhachau, Madnvi, and Gandhinagar in Gujarat, India. See figure below for details.

A. Tiwari, *The Capacity Crisis in Disaster Risk Management*,
Environmental Hazards, DOI 10.1007/978-3-319-09405-2

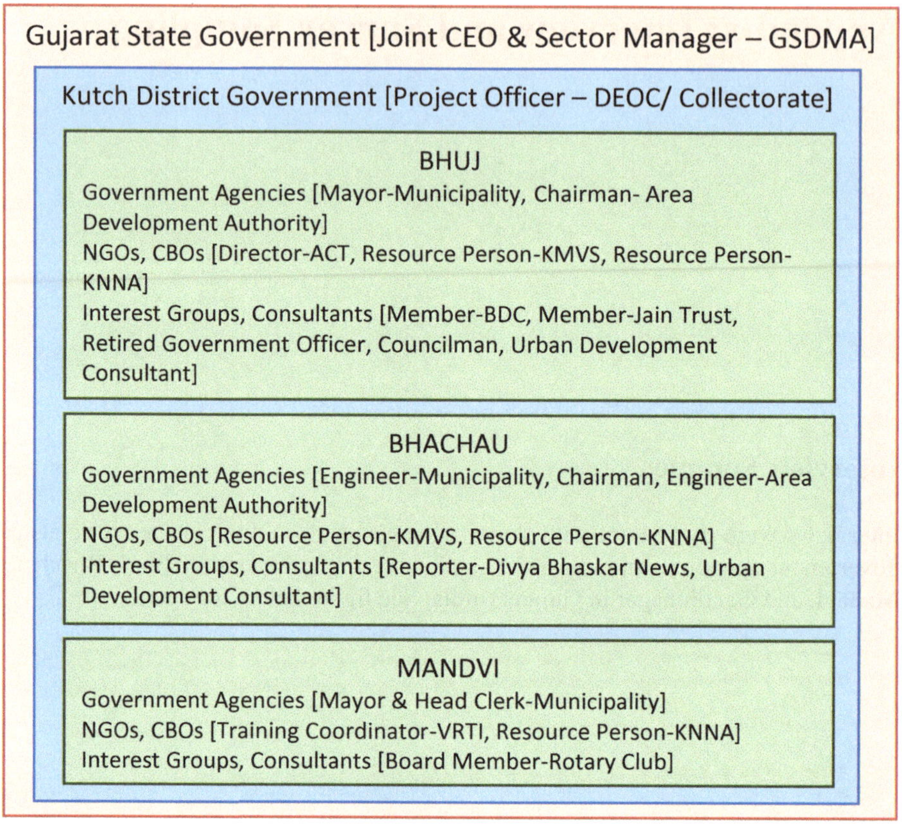

Fig. 1 Interview sample

Survey Sample

Two separate questionnaire surveys targeted government agencies and residents in Bhuj. The surveys in the district and local government agencies targeted the senior staff of the different sections within the agencies (e.g., emergency services, engineering division, and finance). This sample is the same as the population, since the population is small.

The survey also included residents from five different localities in Bhuj who represented different socioeconomic groups (high income to low income) and different disaster reconstruction characteristics (rebuilt at the same place, relocated, slums). These were identified through discussions with the staff at Bhuj municipality and NGOs.

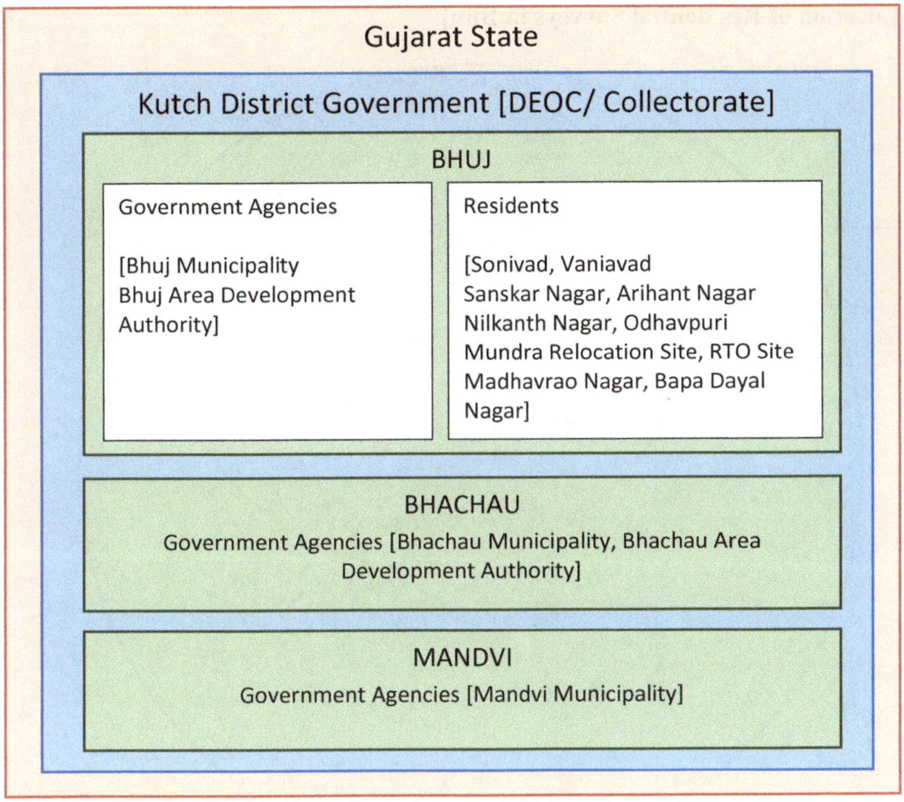

Fig. 2 Survey sample

Location of Residential Surveys in Bhuj

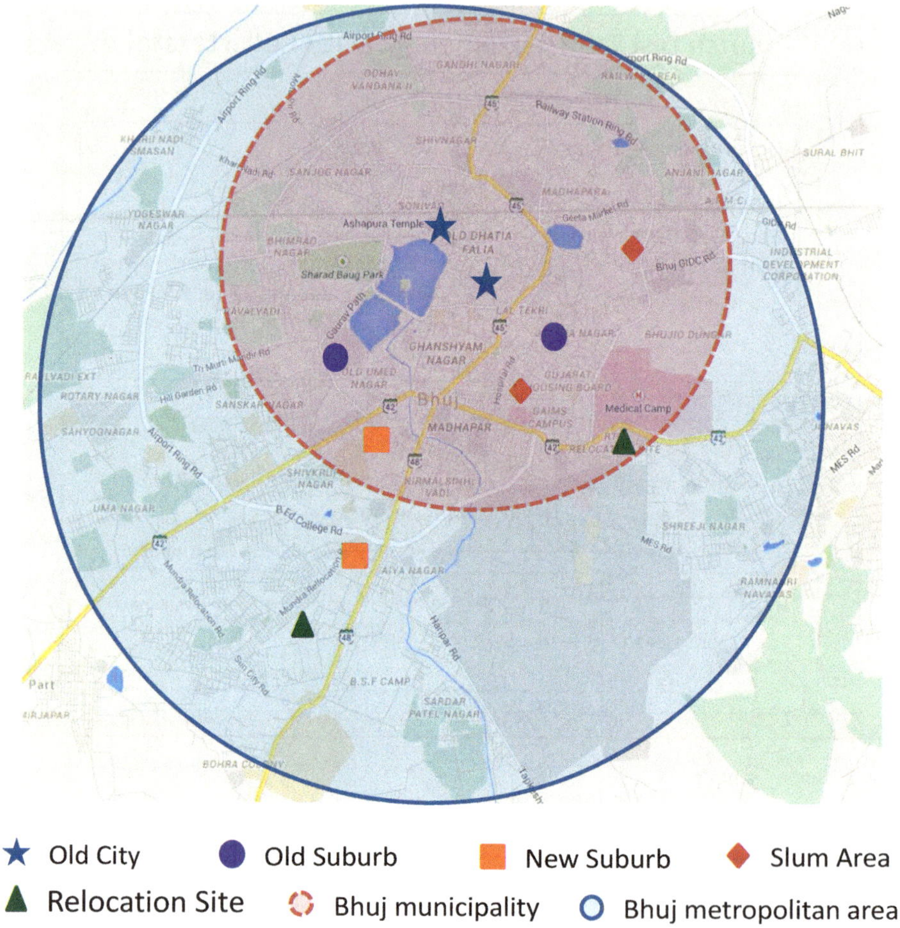

(Map data source: Google, 2015)

Index

A. Tiwari, *The Capacity Crisis in Disaster Risk Management*, Environmental Hazards, DOI 10.1007/978-3-319-09405-2

CPSIA information can be obtained at www.ICGtesting.com
Printed in the USA
LVOW01*1224030515

437052LV00001B/88/P

9 783319 094045